Biological Motion

Biological Motion

A History of Life

Janina Wellmann

Translated by Kate Sturge

ZONE BOOKS · NEW YORK

2024

© 2024 Janina Wellmann

ZONE BOOKS

633 Vanderbilt Street

Brooklyn, NY 11218

Printed in the United States of America.

Distributed by Princeton University Press,

Princeton, New Jersey, and Woodstock, United Kingdom

Library of Congress Cataloging-in-Publication Data

Names: Wellmann, Janina, author. | Sturge, Kate, translator

Title: Biological motion : a history of life / Janina Wellmann ; translated
    by Kate Sturge.

Description: New York : Zone Books, 2024. | Includes bibliographical refer-
    ences and index. | Summary: "Biological motion is the first book to study
    the foundational relationship between motion and life, unearthing the
    long history of motion as the signum of the living world from Aristotle's
    animal soul to contemporary molecular motors". —Provided by publisher.

Identifiers: LCCN 2023013005 | ISBN 9781942130819 (hardcover) |
    ISBN 9781942130826 (ebook)

Subjects: LCSH: Animal mechanics. | Biology—Philosophy. | BISAC:
    SCIENCE / Life Sciences / Biology | MEDICAL / Biotechnology

Classification: LCC QP301 .W38 2024 | DDC 612.7/6 —dc23/eng/20230802

LC record available at https://lccn.loc.gov/2023013005

To Ada and Noga

Contents

Prologue

In Franz Kafka's story "The Metamorphosis," of 1915, commercial traveler Gregor Samsa wakes up one day lying on his suddenly "armor-plated" back and unable to move. His many "pitifully thin" legs "wave helplessly before his eyes."[1]

Getting out of bed is a laborious affair for Gregor. He cannot control his numerous legs, but frees himself by "rocking to and fro" so that he falls out onto the floor. Pushing himself forward, he manages to open the door with the weight of his body. He is still trying to master the motion of a beetle with the mind of a human being and holds himself upright with difficulty by leaning on the door. The chief clerk, sent by the company, flees at the first glimpse of his salesman. When Gregor tries to walk, hoping to explain himself and detain the clerk but "still unaware what powers of movement he possessed," he falls down onto "all his numerous legs." Once on his feet, he feels "a sense of physical comfort" for the first time — his legs are at last "completely obedient."

As Gregor gradually comes to terms with his own movements, the observers of the spectacle experience deep unease at this evidence of his animality. Kafka describes with equal precision Gregor's struggle for motion and the instinctive gestures of repulsion that his insectlike demeanor provokes. The chief clerk backs away without letting Gregor out of his sight, inching toward the door as if "he had burned the sole of his foot" and stretching his arm out to the staircase before rushing down. Gregor's mother springs "all at once to her feet, her arms and fingers outspread," then backs "senselessly away" and sits down abruptly on the table behind her to avoid contact with the floor; his father, brandishing a stick the clerk has left behind, drives him back "hissing and crying 'Shoo!' like a savage" until he pushes Gregor, stuck in the doorway and already injured, back into the bedroom and slams the door.

Now shut in his room, Gregor increasingly accepts the movements of the insect as his own. He begins to take pleasure in "crawling crisscross over the walls and ceiling" and especially in "hanging suspended from the ceiling." Lying down suffocates him, but upside down, he can breathe more freely and entertains himself by falling to the floor, now able to control his body sufficiently to land without harm.

After his first excursion, Gregor is confronted twice more with the vanishing world of human beings — and again it is movements that seal his fate. When his mother and sister come to clear up his room, he rushes out and is injured by splinters of glass. "Harassed by self-reproach and worry," he crawls to and fro until he is exhausted. Then his father returns; a chase ensues. In fear of his life, Gregor runs from his father, "stopping when he stopped," and the two circle the room. Gregor is attacked as an insect, but instead of making use of the repertoire of movement that would enable an insect to escape, he remains deeply attached to his humanness. He fears that climbing up to the ceiling could be perceived as "a piece of peculiar wickedness" and quite forgets that the walls are open to him as an escape route from his father, so superior when it comes to running. His father begins to throw apples at him, and the defenseless Gregor is wounded again.

On the second occasion, Gregor, attracted by the sound of his sister playing the violin, crawls out into the living room, where the family is now sitting with the three lodgers. Injured, weak, dusty, and dirty, he wants to tempt his sister into his room so he can listen to her music alone. This time, we are struck not by the reaction of the others — the three men seem less worried by the insect itself than by the "disgusting" domestic conditions with which they associate it — but by Gregor's decision at last to turn "his frightful appearance" to his own advantage. Using his insectlike mobility and speed, he plans to guard all the doors of his room at once and keep his sister secluded from the world, just as she keeps him secluded. The plan fails. For the first time, Gregor creeps away not backward, but forward. Scuttling feebly under his family's silent gaze, he retreats into his room and dies.

It is the charwoman who finds him on her customary morning peep into his bedroom. Crediting him "with every kind of intelligence," she keeps a safe distance to tickle him with her broom. She is annoyed at his lack of reaction and pokes at him a little harder. When she finds he can be pushed away without resistance, she understands that he is dead. Incapable of any movement at all, Gregor is no longer a human being, nor is he any longer an insect. He is matter: "It's dead, it's lying here dead and done for!"

Nothing Is More Revealing

than Movement

Gregor Samsa's transformation into an insect has exerted much fascination in literary history, but of all the riches that "The Metamorphosis" presents for our interpretation, Kafka's precise description of movements has attracted the least attention.[1] The gradual process by which Gregor becomes aware of his new existence is one of coming to know his new corporeality. In practice, understanding his own transformation largely means assimilating the unaccustomed movements, necessitated by an exoskeleton and articulated legs, of a creature that is described only vaguely, but that clearly belongs to an insect species. The execution of these alien movements is what drives Kafka's story on: the definition and interconnection of the characters by their movements, Gregor's inner reconciliation with his outer form through his new body's movements, and the end of his existence both as insect and as Gregor Samsa at the moment of losing the capacity to move.

The movements that Kafka describes are common ones. Yet almost unnoticed, they form the powerful piers between which the story is suspended — being human, being creature, being alive. Gregor's metamorphosis, which begins with his inner, human, and conscious appropriation of his animal body's movements, concludes with a death that is attested by the external world through his absence of movement.

This book takes as its object of study the triad of motion, animal, and life that Kafka wove into a parable on humanity. My theme, though, is not being human, but being alive, and movement that does not aspire to the human, but descends into life itself. When I quote Martha Graham's motto that "nothing is more revealing than movement," then, I do so not in order to highlight the artfulness of the dancer's (or, indeed, the writer's) choreographed movements. Far more simply, and fundamentally, movement reveals life.

Motion, in this book, is the most profound definition of living existence. It is movement that keeps the living alive, movement that organizes life—from the macroscopic to the microscopic to the molecular level. My investigation is anchored in one of our most constant, least questioned observations on the world around us: a being that moves is a being that lives, and how that living being moves tells us much about what it is.

Biological Motion

The term that forms the title of this book, "biological motion," embraces just that foundational relationship between motion and life. I use the phrase, first, to denominate motion *in* the biological world—that is, in the domain of the physical world that we regard as biological. The biology in "biological motion" is a field of scientific inquiry into the living world that has undergone enormous historical changes. In particular, the science of biology has long since ceased to be one in which knowledge is advanced by "unaided sensory experience." Instead, biological knowledge is driven by "artifice," as the American philosopher of science Nicholas Rescher has put it, in the sense of both artful skill and stratagem.[2] What biology brings forth and legitimates as scientifically relevant objects and questions today differs significantly from the biological world that previous centuries saw: technologies, devices, experimental procedures, and digital practices make life newly visible and newly intelligible, continually expanding the parameters for thinking about it.

But "biological motion" also designates the motion *of* the biological world: movement as the basic characteristic of all things that live. Biological motion in this sense is animate motion, motion within and originating in the organism. The crux of biological motion is that it differs from the motion of physical matter by being active and self-directed. Such motion has been regarded as a sign of life since ancient times.

The fact that I need to introduce the phrase "biological motion" at this stage is itself revealing. The term was coined in the 1970s by the Swedish psychologist Gunnar Johansson to refer to a specific set of experiments in perception that demonstrated the impressive ability of human beings to recognize motion. (I will say more about this later in the book.) I use it here to usher into scientific discourse a phenomenon that is much in need of a name. Indeed, as a conceptual couplet, "biological motion" points to a curious lacuna at the heart of the biological investigation of life: whereas the scientific study of life has prompted much historical research, we know next to nothing about how views of motion evolved alongside it. Put differently, the bond between life and motion in the physical world may be tight, but in scientific studies, the two terms have had unequal appeal, one being as seriously interrogated as the other has been grossly neglected. Yet thinking about biological motion has much to offer. From the perspective of its endurance or transformations, it can illuminate historical shifts in our understanding of life or shed new light on the epistemology of contemporary science. Given that motion bears such enlightening potential, we may ask why it is so manifestly absent from the historiography of life. First, however, comes the question: What is motion in the first place?

On the Science, Knowledge, and Representation of Motion
Motion entered European philosophy, in the fifth century BCE, as an absurdity. In the teaching of Parmenides, there is nothing that is not; in that of Heraclitus, there is only flux.[3] Motion could not exist either with respect to always-already completed and thus inalterable being or with respect to its opposite, the constant transition

and loss of identity of that which moves. Later, Zeno devoted his paradoxes to disproving the existence of motion, whereas Aristotle made motion a cornerstone of his philosophy.[4]

Its epistemic longevity suggests that motion has always enjoyed the privilege of attention, yet motion is by no means one of those concepts that have reliably attracted the interest of philosophers, scientists, or artists (apart from dancers, of course) across the centuries. Though no less indispensable than the notions of space, time, and force, the imaginary expanse of motion does not enjoy a similar status to theirs in intellectual history.[5] The reason may be that motion is not conquered, as spaces may be, does not wither under the dictates of time, does not drive anything except itself. Motion is an unobtrusive magnitude, its essence being to take place without calling attention to itself.

To be sure, many disciplines have tried to throw their snares over motion, with physics leading the way. Motion has always been a fundamental concept in the exact sciences, and physics effectively co-opted motion as an exclusively physical object, framing it with the instruments of geometry and algebra. The early historiography of science, itself largely anchored in the field of physics, located motion's historical and philosophical emergence primarily in the mechanist thinking of the seventeenth century. Where biological disciplines have studied living movement at all, they have virtually never begun to contemplate it outside of those parameters. The criteria were set by the Scientific Revolution, which the history of science has made its founding narrative and nothing less than a "supernova" of the human intellect.[6] Accordingly, they have become the constraints on biology's examination of animate motion while also accounting for the almost total neglect of the issue in the history of biology.

Instead, animate motion remains coterminous with physical motion, whether in the heights of the universe or the depths of the animal domain. Locomotion studies — broadly, the study of how animals, including human beings, move — is thus devoted to animal anatomy and the physical environment in which legs, fins, wings, or cilia are used for movement. Locomotion is the concern of kinematics, biomechanics, and bioengineering or of autonomous motion research on intelligent systems that collapses the boundary between the world of old-fashioned animals and man-made "snail-o-bots."[7]

Where animate motion is not clearly incorporated into the cosmos of physics, it has mainly featured in the science of animal behavior. In a standard definition by the *Encyclopaedia Britannica*, this encompasses "everything animals do, including movement and other activities and underlying mental processes."[8] In the long and diversified history of research into animal behavior, animate motion is subsumed under behavior, absorbed into a general notion of "doing" that places individual movements in the context of stimulus-response and instincts, objectives and intentions, or social interaction and evolution.

The movement behavior of humans, though, is what most of all marks thinking on motion and makes it the object of genuinely multidisciplinary endeavors. In the 1930s, the French anthropologist Marcel Mauss (1872–1950) introduced the notion of "techniques of the body" (*techniques du corps*), founding the French tradition of anthropological and sociological interest in the body. Mauss saw the gestures and other movements of the human body not as something natural, but as a normatively loaded, cultural legacy. In this view, walking, climbing, or jumping—no less than techniques of sleep, techniques of care for the body, or techniques of reproduction—are distinct "ways in which . . . men know how to use their bodies." Being "specific to determinate societies," they demonstrate that not only did non-Europeans swim differently than Europeans did, for instance, but that earlier generations jumped or climbed trees differently than later ones.[9] Since Mauss, the movement of the human body has been parsed in every possible academic direction. Just a few examples are the anthropology of upright gait, the literary figure of the flaneur, and the customs and measured movements of the art of dance;[10] bodily culture from care and crisis, to sports and exercise, to discipline and deprivation;[11] and facial expressions and gestures in physiognomy, theater, and social perception.[12]

But it is not just within and across academic domains that motion is everywhere present and nowhere at home. Constitutively fugitive, motion is hard to portray. It is not a tangible object. Evading perception and representation alike, it slips from being inconspicuous, acting everywhere in the background, into being invisible. Despite this virtual invisibility, however, motion has been present in every epoch of art and picture making and has never ceased to challenge the intellect. Art, technology, and the mind have been taxed by the question of how to capture, freeze, pick motion out of flux, given that ephemerality is its essential property.

Across the centuries, motion has appeared in a movement depicted in a painting and the affects conveyed by that movement, in architecture as measured out by human steps, or more generally in a reality that is accessed bit by bit along the movement of the eye and only successively opens up to perception. Not least, motion in art finds its resonance in the beholders, in the feelings and processes that art provokes in them.[13] This is why Leonardo da Vinci (1452–1519) regarded as "most praiseworthy" the figure that "best expresses through its actions the passion of its mind." The double movement is what makes a representation seem alive. If it fails, Leonardo concludes, the picture seems "twice dead, inasmuch it is dead because it is a depiction, and dead yet again in not exhibiting motion either of the mind or of the body."[14]

Like no other, Leonardo approached "the representation of the *concept* of movement," as opposed to confining himself to the "mere evocation of its physical manifestation."[15] The lively movements he captured on paper arise from a virtuoso use of graphic resources; as art historian Martin Kemp remarks, this wealth of resources was unprecedented and went on to influence virtually every area of graphic art. The repertoire that Leonardo assembled for the first time embraces all the important representational resources available before the invention of moving images.[16]

Starting in the mid-nineteenth century, innovations in optical technology such as flip-books, phenakistoscopes, and chronophotography prepared the human eye to grasp a rapid sequence of images. These new visual habits paved the way for science to deploy an entirely novel set of possibilities of representing nature, and the new media opened up fresh access to motion. Cinematography, video, motion capture, and the computer are all successor technologies that enabled movement to be visually generated as a reproducible event beyond its immediate execution. In recent decades, they have attracted a great deal of attention in art, philosophy, and the history, theory, and archaeology of media, and as the computer has come to prevail as a pictorial tool, "animation" has become a transdisciplinary buzzword in research. Now that computer-generated and computer-animated images dominate our visual culture, their aura of aliveness intrigues those disciplines as the apparently natural obverse of their technicality.[17]

Even the most artful of such movements, however, are bound to remain only representations. This is true of artistic representation in pictures, buildings, and sculptures, but it also applies to the manifestation of motion in equations, notations, films, or computer programs. Descriptions of motion are not and cannot be motion itself; they serve at best as imitations, analyses, or instructions on how to reproduce it as event.

Biology on the Move

All this gives us little guidance in our exploration of movement in the living world. Yet a book about biological motion seems timely, for a glance at scientific journals and websites suffices to show that motion is a crucial or even the crucial theme of basic biological research in the twenty-first century. At the very beginning of the millennium, the journal *Science* proposed, not without dramatic flourish, that movement is the "root of all existence."[18]

Far-reaching advances in visualization technology, fluorescent marking, and the rise of systems biology have meant that processes deep in the body's interior, at a cellular and subcellular level, can now be made visible. A new iconography of the body has taken the place of the external gaze, shaped for centuries by the practices of anatomy, that intrudes into a body dismembered and denuded. Now the body intact, revealing itself from the inside out, guides the gaze as a participant through the darkness of its own interiority.

In this body, nothing is still; everything moves. Indeed, the body's task seems to consist solely in sustaining motion. Pivotal physiological processes are now studied by scientists everywhere in terms of the movements they make possible: the metastasis of cancerous cells, the migration of axons, the movement of neutrophils to the site of injury, the migration of cells during ontogenesis, intracellular transport, the walk of motor proteins in the cytoplasm. Scientists and lay people alike take delight in picturing how "a migrating cancer cell trails sticky appendages as it rolls through a blood vessel and attempts to squeeze through the vessel wall" or "a fiery orange immune cell wriggles madly through a zebrafish's ear while scooping up blue sugar particles along the way."[19]

It is surely impossible to deny the special relevance of motion for biology today. But how exactly does the movement of proteins, leukocytes, or tumor cells take place? When we talk of sperm that "swim" and immune cells that "roll" or "wriggle," does the nuanced language used to describe movements define them scientifically, or only metaphorically? Historically, how did this descent into ever greater depths of the organism's interior result in the unstoppable ascent of motion as a concept? Finally, is the current pertinence of the topic really new?

Siting Movement

My project began from the observation that watching the movements of living organisms is among the most mundane and self-evident ways in which human beings assure themselves that they and the other members of their world are alive. Movement is not an object to be found: it happens. As a phenomenon manifesting itself in continual change and momentary consummation, it is performance and reiteration that gives shape to movement. Movement thus occurs within a situated interplay of place, time, and viewer.

This book, first, examines how biological motion is produced in particular scientific interplays of conceiving, perceiving, and producing biological motion. It offers signposts to mark the sites where researchers, technologies, ideas, and practices set off on new paths in order to constitute the phenomenon of motion.

Drawing inspiration from art history, I describe the situated interplay in which motion is made as a "site." The art historian Peter Gillgren has given a very valuable definition of the site as a "conceptualized place with strong internal and external relationships" and "explicit openness of meaning."[20] In the 1960s, artists started to make art specifically for a site and its context in the natural environment, the urban cityscape, or the museum.[21] Art historians have studied such artworks by asking how the beholder's movement and perception engenders the artwork at its specific location as an ensemble of place, work, and viewer. Today, that method is no longer confined to art of the modern era, but is applied much more broadly to artists and their works.[22]

The concept of the site brings to the fore the constantly renewed perceptual, cognitive, and participatory dimensions of art as an event. Siting involves attentiveness to perception of place and circumstance with all the senses, by all actors in their own moments of time. The scientific investigation of biological motion is likewise a specifically situated interplay. It entails observers encountering a sensory event that usually they themselves produce as an event, using the instruments of analysis, experiment, or computers under the mandate of a particular scientific question. By capturing motion visually or as data, they make it into an object. The phenomenon, the beholders, and their instruments and data form an ensemble that only as such can constitute "motion." Methodologically, thus, siting movement expands the investigation of movement, bringing into play the confluence of the fleeting event's sensuality, its perception, and its intellectual analysis that together make up the motion event. Siting thus highlights, on the one hand, the changing external relationships that constitute motion, thwarting complete-ness and closure. On the other, it treats motion, like the artwork, as independent in itself, a regulated, orchestrated interplay and sov-ereign pattern of strong internal coherence. It is this duality, I hold, that biological motion shares with a site-specific work of art: self-contained, yet existing only in relation to its environment, being simultaneously in itself and in the world.[23]

Second, the concept of site allows me to look at the biological organism itself as a site of movement and the organism's dynamics as an interplay between motion and environment. The processes that constitute the organism, keep it alive, and renew it again and again are always movements, but they are produced by particular situations, in specific ways, in diverse micromilieus, each with its own constraints. Never static, the physiology of the organism is what motion makes of it, new at every moment. If, as Nicholas Rescher put it, events have little or no "fixed nature in themselves," neither do their effects.[24] Observing a movement event, and thus the workings of the body, will always be only a snapshot, a momen-tary impression in every sense. It is momentary for the viewer who participates in it; it is a momentary excerpt from the flow of motion, a momentary instantiation within a specific experimental setting, and a momentary use of the artifices available to snatch movement out of ephemerality.

Elusive and protean, motion is difficult to pin down. Its story resists narrative. It cannot be told in terms of an ending and con-clusive analysis, as narrative necessarily is. Instead, and this is my third point, this book makes siting its method in following bio-logical motion through history. I pursue movement and the arrival of thinking about biological motion on the historical stage by means of movement's own devices — by striking out on a journey. Journeys generate events with every step; they construe motion out of motion. Siting is just such a process of "gradually be[ing] choreographed into learning about the site." As Peter Gillgren has elegantly described it, "with every step, new features and new constellations appear."[25] In the same way, this book identifies historical moments when thinking on animate motion takes new turnings, steps off established paths, steps onto new ones. As it moves along its route, the book notes some of the waymarks that locate biological motion at the intersections of knowledge domains and scientific and cultural practices—the animal machine, modeling in mathematics, or the human gait in the chemistry of molecules.

These movement events are neither unalterable nor arbitrary, but the outcomes of my own, mobile positioning. They form a choreography of the phenomenon of biological movement as, both historically and analytically, it glides between the most varied fields of knowledge, regroups at the boundaries of questions and methodologies, shape shifts along with the technical media that bring it forth and the experiments that make it visible. Motion thwarts not only closure, but also the interpretive power of narrow, static approaches, traditional paradigms, and established conventions.

Movement, in other words, shows us the "mental sculpting" (to borrow Martin Kemp's term for Leonardo da Vinci's method of drawing with its multiple revisions) that is performed by our own labor of thinking.[26] For the confrontation with motion is always also a confrontation with the observer's own movements of perception and understanding. Movement challenges the observer to engage with the phenomenon rather than pinning it down, to seek rather than to find, and to give form to something that is unfinished and inaccessible.

Stepping into the Book

Attention to life moving on screens may be a recent phenomenon, but moving matter never escaped notice or scrutiny. On the contrary, it was always part and parcel of the scientific study of life and had profound epistemic consequences for the whole of biology from the start. It has been the driving force for research that seeks the foundations of life, or at least the foundations of its own knowledge. *Biological Motion* uncovers that secret life of movement — eagerly explored throughout the centuries by a veritable crowd of monks and microscopists, botanists and bacteriologists, taxonomists and cytologists while hidden from historiography in plain sight.

Animate motion entered the microscopic world as a mystery, captivated Enlightenment audiences as a curiosity, overturned the notion of life in the nineteenth century, and has recently become the key to framing molecular existence. Over the centuries, the locus of animate movement migrated from animal to matter, from the organized to the inchoate, from the organ of locomotion to the contractility of all living matter. Descending from the whole to the part, from outside to inside, movement became the determining feature and essential explanation of the inner workings of life, whether reproduction, physiology, or protein action.

In the seventeenth century, Antoni van Leeuwenhoek's discovery of animalcula in a drop of rainwater — tiny animals that became visible only under the strong magnification of his lens — opened up a new realm for biology, which would only much later begin to recognize its true luxuriance. In the visible world, motion was what was obvious; with the discovery of movement in the microscopic world, motion was what was extraordinary. Yet Leeuwenhoek entertained not the slightest doubt that this new world was full of life.

But tiny animals with unlikely behavior were only the beginning of an investigation that by the nineteenth century had shaken some of science's most deeply rooted beliefs about the organic world. Simple as they were, infusoria presented a series of mind-boggling puzzles. First of all, they posed the question of what constitutes an organism. If infusoria moved like animals, and consequently were animals, one might expect that on closer inspection and experimental scrutiny, they would prove to be organized with similar complexity. If they did not, would they still be animals?

The self-evident analogy, based on motion, between higher animals and infusoria began to crumble in the nineteenth century.

Motion took on a new role as the criterion for redrawing the frontiers between animal and cell, plant and animal, dead and alive. In protozoology, taxonomy, cell biology, botany, and physiology, animate motion expanded from being the signum of the animal to defining protozoa and plants, then organic matter itself. By the end of the century, the activity or contractility of organic substance, and this alone, was what indicated the property of being alive. Yet in order to track down organic matter's contractility and understand how all the various physiological functions of growth, nutrition, and development arise from, work through, and are perpetuated in motion, a world full of vitality had to be studied while it was still alive.

The microcosm glimpsed through the lens faced its beholders with visual and epistemic challenges that lost none of their appeal for the curious-minded over time. The effortlessness with which life exudes vitality is matched only by the arduousness of biology's search for ways to bring the hidden life of biological motion into the realm of perception and analysis, its struggle with the practices, methods, and devices it must employ.

Even in Leeuwenhoek's day, projections using light and lenses offered a space of experimentation for a new, sensual experience of microscopic life as it lived. Performances and spectacles featuring the camera obscura, magic lantern, or solar microscope were well known, as was their potential for studying movement. In the seventeenth and eighteenth centuries, watching insects and infusoria was not only an intellectual challenge, but a pastime. In the nineteenth century, solar microscopes and special illumination methods were still being used to make motion visible and thus open to investigation, even if the motion was so delicate it could hardly be perceived or else took place in material that was itself almost invisible and formless.

Well into the twentieth century, scientists continued to work on a broad repertoire of inscription formats and visual methodologies that enabled movements to be recorded and measured successively on paper, played forward or backward on celluloid film, or studied three-dimensionally and back-to-front through prisms—at times in a resolution high enough to register the individual cell.

The discovery of vitality that arose from microscopic exploration brought with it the wonderment, unbroken for centuries, at how perfectly mysterious this world in motion is. Leeuwenhoek's tiny organisms were animals by virtue of their movement, but the nature of that movement remained difficult to grasp. Some characterized it as spontaneous, voluntary, instantly recognizable; others likened it to acting on the stage and found in it a spectacle equal to anything that the opera and street performance could offer the pampered eighteenth-century Parisian theatergoer. To no small extent, the epistemic challenge of understanding moving life—fleeting but vital, unremarkable but fundamental, ubiquitous but easily overlooked—has been a source of aesthetic pleasure. However much it is parceled and partitioned into ever tighter scientific grids, movement has never lost its imaginative allure. On the contrary, the modern life sciences seem to have fallen entirely under its spell.

As research has pried more deeply into the subcellular domain and microscopy has become nanoscopy in the past few decades, enormously sophisticated experimental apparatuses and biochemical knowledge are paired with the computer and mathematical modeling. While animalcules wriggled visibly beneath the seventeenth-century microscope, the new microscopy also sees cells, nanobots, and proteins that walk or limp. When single molecules are drawn into the visible world, it takes mathematics to make the invisible move under the conditions of visibility. Now that motion is visible in new places, is it also visible in new ways that help us to understand it differently?

Aristotle established motion as a fundamental category for thinking about, perceiving, and organizing the world. The body is alive because the soul lends it life, and living existence maintains itself solely in movement—the precondition for its constant transition and transformation. Motion in Aristotle's work projects organic existence toward its future, its preservation in mutability. From this intellectually powerful enigma, motion descended into the more prosaic conceptual space of the machine in the seventeenth century. Now animals moved according to the same rules as projectiles or planets, whether in Pisa or in the firmament, and machines could copy and perform motion in the place of animals. In the twentieth century, technical experimentation relocated motion into the perceiving observer, and movement became a phenomenon of ascription. Today, synthetic robots and biohybrids move organically in ways that play with our perception and erode the Aristotelian equation of being in motion with being alive.

The critical question we face in the twenty-first century is no longer whether what we see moving is natural or an artifact or both, but the inverse: whether or not we wish to define the motion we see as animate. Does biology's mathematical turn finally liberate biology from the problem of how to explain aliveness or, quite the contrary, does it strip life of its very essence, its vitality?

CHAPTER ONE

What Itself Moves Itself

Movement is ubiquitous, in the heavens and on earth. Stars move, and so do living beings; particles of matter move, and so do an animal's legs, wings, or heart. The nature of that movement had puzzled mythology, philosophy, art, and manufacturing ever since ancient times, but seventeenth-century physics, mathematics, and mechanical philosophy finally declared the question resolved. In doing so, they opened up new horizons for physics, but simultaneously closed down worlds for biology. The physics of locomotion fenced off a scientific domain that more or less excluded from inquiry what was the most profound biological conundrum.

Because of its orientation toward physics, the historiography of science has obscured a wealth of debates, dating back to antiquity, about the complexities, inconsistencies, or outright impossibility of understanding motion in the living world. In fact, motion continued to test the limits of thought and called for more embracive approaches than that of physics alone. The fascination with motion comprehended everything from amphibians to the heavens via analytic geometry and lens grinding, philosophy and experimental practices—pursuits that sometimes converged in a single person. At the origin of that fascination was just one fundamental observation, that only living beings move of their own accord. Yet a staggering lineup of models and ideas was assembled and freely recombined in search of an answer to the question: Why?

Motion Is Change

For Aristotle (384–322 BCE), movement inheres in everything natural. Motion is a fundamental fact of our experience of the world. As such, its discussion forms part of the *Physics*, the collection of manuscripts in the fourth century BCE *corpus Aristotelicum* that deals with the general principles of natural things — not only animals and plants, but also the "simple bodies," earth, fire, air, and water (*Phys.* 2.1.192b10).[1]

In the *Physics*, motion characterizes all natural things, to each of which Aristotle attributes "*within itself* a principle of motion and of stationariness." Importantly, motion means every kind of change. It is all-encompassing and takes four different forms: as change with respect to quantity (growth or decrease), quality (white or black), substance (coming to be or passing away), or place (upward and downward or movement from one place to another) (*Phys.* 3.1.201a3–9). In other words, Aristotle's view of motion covers everything in nature that is processual (*Phys.* 7.2.243a13–14).

Aristotle defines motion as "*the fulfilment of what exists potentially, in so far as it exists potentially* ... of what is alterable *qua* alterable" (*Phys.* 3.1.201a10–11).[2] Motion in this definition has a starting state and an ending state; more precisely, it is a tension between potentiality (*dunamis*) and actuality (*energeia, entelecheia*). The two terms that Aristotle uses to denote actuality, *energeia* and *entelecheia*, both of which he introduced into ancient Greek himself, are applied interchangeably in his work, but highlight the contradiction contained in the concept of actuality: *energeia* means — in Joe Sachs's translation — "being-at-work," *entelecheia*, "being-at-an-end."[3] Actuality means both "to be at work" and "to act for an end" and thus is a movement between the possible and the actual, which are two elements of one and the same motion. It is, but equally is not yet; it is directed to a future.[4] Being real, being actualized, therefore means a vital being-in-the-world and a being-on-the-way-to-completion, which is always an active being, directed toward self-maintenance or one's role as part of a greater whole.[5]

Motion, recall, inheres in all natural things, and natural things are epitomized by living beings. "Life," as Aristotle explains in *De anima*, has "more than one sense, and provided that any one alone of these is found in a thing we say that thing is living. Living, that is, may mean thinking or perception or local movement and rest, or movement in the sense of nutrition, decay and growth. Hence we think of plants also as living" (*DA* 2.2.413a22–25).[6] In this expansive sense of motion, as every kind of change, all the properties of life are changes in quantity, quality, substance, or place, and thus movements. In line with his predecessors (*DA* 1.2.403b20–404b), Aristotle also defines motion as the principle of the soul, because the soul *is* motion. The soul bestows motion, and thus life, because it "is in some sense the principle of animal life" (*DA* 1.1.402a6–8). To be alive is to be animated — and to be in motion.

Not everything that is alive possesses all of the capacities that belong to life. Aristotle distinguishes three powers of the soul, each granted to different life forms: the vegetative (responsible for growth, metabolism, reproduction; possessed by plants), the sensitive (sensory perceptions; possessed by animals along with the vegetative soul), and the rational, possessed by human beings alone (*DA* 2.2.413b11). With these powers, the soul drives the future-oriented movement from a life that is potential to an actualized life. In this sense, motion includes the sustenance and preservation of the living being, ensuring its continuity and identity across its constant change. Even plants, sessile as they are, nourish themselves, come to be, and pass away — so they, too, are alive, animated, and consequently in motion even though they do not locomote.

Aristotle's *De motu animalium*, a kind of sequel to *De anima*, is less a simple appendage to the earlier work than an attempt to "perfect the account of the soul."[7] It turns very concretely to the questions of "how the soul moves the body, and what is the origin of movement in a living creature" (*MA* 6.700b10).[8]

How the Soul Moves the Animal

Aristotle fundamentally thinks of the soul as relating to a concrete, living body. The soul is the principle of the body's motion, a principle that is specifically realized in the body. In *De motu animalium*, Aristotle therefore connects his study of how the soul acts in living beings with a consideration of the anatomical and physiological constitution that makes them capable of carrying out movements in general. To be sure, *De motu animalium* investigates the physiological conditions of movement solely for the case of animal locomotion. Because locomotion begins with a perception, it is a power reserved to animals as the possessors of a sensitive soul.[9]

For Aristotle, the physiological organization of locomotion begins in the animal's heart. The heart is fundamental to Aristotelian philosophy, the center of the corporeal organization of living beings. Perception happens when perceptual alterations are transmitted to the heart from the body's periphery (the sensory organs). By its nature, the animal is organized to react to stimuli and perceptions from its surroundings in a way that enables it to preserve itself and provide for its own physical well-being — that is, by striving for agreeable feelings (as in nutrition) and by seeking to avoid disagreeable ones (as in pain). Pleasant and painful feelings induce thermal alterations (heating and cooling, respectively), which are of special importance for animal motion. The thermal changes result in the contraction or expansion of the *pneuma*. All living beings possess this "connatural spirit," from which they "derive power" (*MA* 10.703a10). The *pneuma* effects movement mechanically, through a "thrusting and pulling" that is driven by its expansion and contraction (*MA* 10.703a20–22). Aristotle considers it obvious (*MA* 10.703a9) that all beings have *pneuma*, some kind of connate air, and tells us that it is to be found near the heart, heavier than fire and lighter than earth (*MA* 10.703a24–25), yet it remains unclear precisely what the *pneuma* is. Aristotle is far more interested in its function, which is to convert the qualitative thermal alterations into mechanical movements of pushing and pulling.[10]

The question remains of exactly how the *pneuma* causes the limbs to move. Aristotle compares the mechanics of the human locomotor system with that of an automaton, proposing a correspondence between marionettes and animals as regards the various elements involved in their motion:

> The movements of animals may be compared with those of automatic puppets, which are set going on the occasion of a tiny movement; the levers are released, and strike the twisted strings against one another.... Animals have parts of a similar kind, their organs, the sinewy tendons [i.e., *neura*] to wit and the bones; the bones are like the wooden levers in the automaton, and the iron; the tendons are like the strings, for when these are tightened or released movement begins. (*MA* 7.701b2–10)

Aristotle describes the *neura* moving the animal in analogy to the strings moving the puppet.[11] These *neura* or sinews as Aristotle understands them are not identical with muscles, but encompass what modern anatomy sees as distinct sets of structures—ligaments, tendons, nerves.[12]

The contraction or expansion of the *pneuma* moves the *neura* in the heart by pushing or pulling them. Aristotle does not explain how the "tiny movement" of the heart's *neura* is then transmitted to the periphery of the body and is somehow amplified on the way. It is clear, though, that he is thinking of some kind of causal chain: as the *pneuma* contracts and expands, the heart's *neura* relax or tighten (*MA* 10.703a18–23) and transmit that motion to other *neura* in the body, which move the bones and thereby the limbs. In concrete terms, motion occurs when something moved responds to something unmoved: "each animal as a whole must have within itself a point at rest, whence will be the origin of that which is moved, and supporting itself upon which it will be moved both as a complete whole and in its members" (*MA* 1.698b5–9). What is unmoved is the support that serves the limbs as resistance, a fixed point. Within the body, the unmoved part might be a joint, for example — for the *neura* connect the bones in such a way that one bone can always move in relation to another, unmoved bone.[13]

The etiologically oriented among Aristotle's zoological studies, especially *De incessu animalium* and *De partibus animalium*, continue the empirical discussion of specific forms of beings and their movements, whether swimming, crawling, or flying.[14] These works are the first in classical Greek philosophy to show that bodily structure and environment determine the type of motion physically and physiologically. Human beings and animals equally are framed within a larger whole of which they are only parts.

Aside from the bodily organization, Aristotle describes the locomotion of animals as driven by "intellect, imagination, purpose, wish and appetite. And all these are reducible to mind and desire" (*MA* 6.700b18–19). Animal locomotion, then, is stimulated by the environment and perception and fostered by desires and more complex intentionalities. This leaves quite some room for debate on the concept of self-motion. Animals can initiate movement on their own, but they may be prompted by influences from the environment, sensory impressions originating outside themselves, and feelings. That is further complicated by the role of the higher cognitive faculties, imagination and intentions, on the basis of which living beings exercise locomotion directed to an end.[15] Finally, Aristotle attributes self-motion not to living things alone, but to all the things of nature (*Phys.* 2.1.192b12).

Scholars disagree as to whether animal self-motion can be successfully inserted into the general theory of motion set out in the *Physics*, which anchors the final justification of motion in a final unmoved mover — an infinitely regressive explanation.[16] Klaus Corcilius, for example, regards the self-motion of animals due to their soul as compatible with the general system of motion assembled in the *Physics*, since the soul is the "unmoved mover of the body."[17] In fact, it is disputable whether Aristotle even has a rigorous formulation of self-motion at all. Self-motion in a strict sense,

Corcilius argues, is not to be found in Aristotle's work — either in his physics or in his notion of living beings. Wolfgang Wieland tries to resolve the dilemma by reading self-motion as a "moved self-moving" in which an external cause is always also possible, such as perception for animals or environmental influences for growth processes.[18] Leaving aside the intricacies of the Aristotelian philosophical edifice, however, what is most significant for my argument here are the parameters that Aristotle opened up for motion.

Everything that lives, including the animal, has a soul, and the English word "animal" still carries that soul (*anima*) in its etymology. At the same time, Aristotle shows us an animal whose movement, although a soul is responsible for it, is also a function of interlocking physical functions and the expression of particular perceptions or complex motives. In the animal, the discernible aspects of motion coincide with something far more impenetrable. Animal locomotion is the segment within Aristotle's spectrum of physiological change where motion is literally graspable. Locomotion presents itself for investigation — as imitation and copy; as reproduction in technical artifacts, automata, and machines; in the interplay of muscles, nerves, and thermal changes.

Figurations of Artificial Motion

Locomotion, Aristotle explains in the *Physics*, is the only form of motion that can be changed from the outside by means of artificial action, specifically through the action of "pulling, pushing, carrying, and twirling" (*Phys.* 7.2.243a15–19).[19] For that reason, locomotion can also be artificially produced by means of *techne*.

In antiquity, *techne* embraced both art and technology, a subordinate form of art, it did not have the status required to bestow knowledge. Articulated puppets like the ones that Aristotle describes already existed in ancient Egypt. As for Western culture, the *Iliad* (seventh to eighth century BCE) tells of artificial helpers created by Hephaestus, the god of fire and blacksmiths, that moved as if of their own accord on golden wheels. According to Hesiod, Hephaestus also fashioned the figure of Pandora out of clay as a present for Epimetheus. Pandora enchanted all with her irresistible beauty and gifts, yet by emptying her box, she brought evil into the world.

The automaton seems to move of its own accord (the Greek word means "self-mover"), an impossible feat, since such movement is reserved to animals alone. Coupled with its maximally close similarity in external form to the animated being, this gives the automaton a wondrous quality, whether in myth or in the complex artifacts devised by Heron of Alexandria as early as the first century BCE. The automaton is an illusion of life and as such, Hans Blumenberg once wrote, a "cunning maneuver or 'machination'": a product of human beings' artistry, but equally the proof of their susceptibility to beguiling deceits.[20]

Artifacts of a motion to which they had no right, automata were ascribed shifting forms of agency throughout the European Middle Ages and Renaissance — not always a divine agency, but generally expressing some kind of magical or preternatural power. Automata came to the Latin West in the ninth century from foreign, non-Christian territories and presented an unfamiliar knowledge, lost to the Latin language since antiquity. They were built by European artisans from the mid-thirteenth century on. Populating churches from the mid-fourteenth century, mechanical Christs distributing blessings, copper angels gliding through mechanical heavenly spheres, and mechanical performances of biblical scenes became routine elements of religious practice — a representation of divinity in the machine that the reformationists of the fifteenth and sixteenth centuries would find uncanny. As elaborate clocks and water features in royal parks, aristocratic palaces, and grottoes, Renaissance automata also had a worldly presence. They expressed the most sophisticated artisanship available at the time and represented political power through entertainment and amazement.[21]

But aside from their virtuoso craft, bemusing beauty, and wondrous operation, mechanical devices are built for a very practical purpose: to relieve human beings of movement and to perform labor. Any movement that can be explained mechanically is in principle reproducible by a machine; in a machine, writes Georges Canguilhem, "movement is a function of the assemblage, and mechanism is a function of configuration." A machine, then, is an artificial construct whose "essential function depends on mechanisms." And a mechanism, in turn, is an implementation of movements, "a configuration of solids in motion," an "assemblage of deformable parts."[22]

Mechanical motion turned machines to human applications. Renaissance engineers built devices able to perform the movements that were too strenuous for humans and animals, as well as creating new movements quite beyond the reach of human and animal strength and physiology. Leonardo da Vinci's drawings imagined every possible and impossible movement machine for land, water, and air, while the sumptuously illustrated machine books or *Theatra machinarum* of the sixteenth century onward — abundantly translated, reprinted, and plagiarized — documented ever more ingenious mechanisms for driving mills, looms, or presses. This genre of engineering treatise, beginning with Jacques Besson's *Théâtre des instrumens mathématiques et méchaniques* of 1578 and Agostino Ramelli's *Le diverse et artificiose machine* of 1588, culminated in epoch-making enterprises such as the hundred-volume *Description des arts et métiers* collated by the Académie des sciences in the late seventeenth century and Diderot and d'Alembert's *Encyclopédie* in the eighteenth.[23]

Motion delegated to machines is a powerful tool to transform human action on the planet. The movements of machines aided the rise of manufacturing and global trade in the seventeenth and eighteenth centuries, and their destructive potential defined the battles of the Thirty Years' War, which plunged the continent into devastation and its inhabitants into existential crisis.[24] As calculators, the filigree product of mathematical and artisanal ingenuity, mechanical helpers would contribute to the development of early modern government, commerce, and in the long run, our ideas about reasoning with machines.[25]

When René Descartes (1596–1659), philosophy's foremost advocate of mechanism, presented his *bête-machine* in the 1640s, therefore, it was not the conceptual company kept by man and machine that was new. Nor was his radical reinterpretation of the animal in the "animal-machine" as momentous as the extravagant interpretive power he now granted the machine. In the seventeenth century, the machine changed from being an automaton of dubious (if convenient) capacities and agency to being "the vehicle of universal transparency." It became, Alexander Sutter continues, the crucial means of coming to know nature. The object of natural knowledge was now the machine itself.[26]

Such mechanist thinking depends upon a peculiar logic by which the machine, though built in imitation of a living model, does not copy that living model, but itself comes to model the natural phenomenon. The machine is not the copy of a body according to which it has been built; on the contrary, the copy is the blueprint for the body that it originally copied. This convoluted figure works only because the machine, a material construct, takes on a clearly epistemic task: in the machine, the cognitive faculty of human beings is contraposed with an organic nature of which the machine itself is not part. And the rationality manifested in the machine may not be part of nature, but once nature is taken apart by reason, it requires reason's order.[27] That has enormous consequences for the figuration of biological motion. Once the machine is the sole point of access to living motion, its imitation of movement becomes one with knowledge of movement.

But if the animal-machine was to wreck the Aristotelian notion of the soul conceptually, Descartes needed to incorporate *all* physiological motion — not only locomotion — into the machine, and that required the complete separation of *res extensa* (corporeal matter) and *res cogitans* (mental substance). Seventeenth-century physics thought of matter as purely passive in itself, to be set in motion only by external forces. Everything that exists and occurs in the world can be explained by the collision of particles of matter in space.[28] For Descartes, that principle applies in equal measure to all organic life. The organic body, whether animal or human, and the entire remainder of the organic world—its generation and decay, its sensibility and perception, and all its vegetative and noncognitive qualities—are fully explicable by corporeal matter (*res extensa*, extended substance) with its particles and movements. Just as for other matter, the movement of organisms can be explained entirely by material conditions. Nonhuman animals, here, fall into the same category as machines, and Descartes excepts only the human being, in that humans alone possess the ability to think. Thought is a property not of matter, but of the *res cogitans*.

Accordingly, Descartes attributes all of the animal's living movements to an entirely mechanical principle, which he locates in the heat of the blood. This heat is the result of particles of matter colliding. It does not differ in any way from the heat that may be emanated by other substances: the fire of the heart, a *feu sans lumière*, arises like every other fire from violent collisions between small bodies hurtling through space. In Descartes's physics and physiology alike, mechanical collisions of particles are the sole cause of motion. The only difference is that in the case of organic bodies, it takes place inside the blood, where the heat produced ensures that the blood expands in the heart's ventricles, enters the vessels, and is transported through the body, keeping the blood in circulation.

Unlike William Harvey, who had described the circulation of the blood for the first time in 1628, Descartes ascribes no role to the contraction of the heart as a muscle.[29] Instead, all the organic processes are regulated solely by the heat produced in the heart: "the heat in the heart is like the great spring or principle responsible for all the movements occurring in the machine."[30] In his account, Descartes breaks up the different organic processes into mechanisms and subsystems of nutrition, growth, generation, and the circulation of blood and adduces various machines: hydraulic automata to explain the circulation of blood, the church organ and its pipes to explain the nervous system, the clock as a universal regulator of tempo.[31] Attributed to the collision of particles and organized into isolated mechanisms, motion is locked inside the organism. Just as his notion of the *res cogitans* extricated thought from the human being as body and organism, Descartes here extricates the organism from its environment. Once set in motion, the body itself creates the conditions for its continuing to move.

Now that only mechanical motion or locomotion was left in the physical and organic world, Descartes helped to supply the analytical and mathematical tools required to quantitize it. Galileo Galilei (1564–1642) had begun parsing movement into space and time experimentally when he measured movement by rolling a sphere down an incline and correlating points on the route traveled by the sphere with the time it took to reach the bottom.[32] It was to solve similar geometrical problems of space-time correlation by means of calculation that Descartes developed analytic geometry. In his system, points in space and time were uniquely definable and quantitatively correlated, precisely measured in coordinates. Motion was visually represented as a moving figure in the coordinate space of a graph between the x axis and the y axis.[33] Toward the end of the seventeenth century, the infinitesimal calculus developed by Isaac Newton (1642–1726/7) and Gottfried Wilhelm Leibniz (1646–1716) then supplied a mathematical formulation of the relationship between the discrete and the continuous.[34] The mathematical tools and methods were now at hand that continue to describe motion today.

It took centuries to move from mythology to machines, from soul to reason, from cunning maneuver to mathematical knowledge. But no matter how the machine is designed and its motion measured, it will not move spontaneously. A machine maintains motion, but the first movement remains the result of a creative act and design, the transcendental principle — a "snap of the divine fingers," as François Duchesneau has put it. Theology formed the hidden, yet necessary Other of mechanical philosophy and of the comprehensive mechanization of nature in the seventeenth century.[35] By contrast, anatomy remained a faithful ally of the soul well into the eighteenth century. To confront the stubborn resistance of life to thought, anatomists favored a robustly practical intrusion into the organism using knives, lenses, and needles. Whereas the figurations of mechanical motion successfully concealed the fact that they had never managed to dethrone the soul, anatomists examined the soul in action. Yet the deeper they penetrated into the tissue and organs of the body, the more they eroded the Aristotelian notion of an embodied soul.

Experimenting on the Soul

In *De motu animalium*, Aristotle had examined the workings of the soul embodied in animals based on his own dissections and vivisection of their bodies.[36] He identified several bodily structures, the most significant being the *pneuma*, heart, and sinews. As noted above, the heart occupies the central position: it is the "origin, principle or seat" of movement and the soul in the body.[37] Whereas the heart itself is passive, the innate *pneuma* that it contains is capable of contraction and expansion. The soul, the unmoved mover, imposes movement on the *pneuma*, which thus acts as "embodiment" of the soul, as the "kinetic force" in the chain of events which results in movement.[38] Originating in the heart, motion is then transmitted to the limbs via the sinews, or *neura* in Aristotle's terms.

These *neura* perform three different roles in the process of movement, which today are assigned to different anatomical structures: they act as nerves (in the sense of transmitters), as ligaments and tendons (binding the joints together), and as muscles (tightening and relaxing, thus bending and stretching the limbs).[39] In other words, the *neura* are the *anatomical* structures of ligaments, tendons, and nerves, but they have the *function* of a muscle: contraction and relaxation. The anatomical structure that we today call muscle, by contrast, Aristotle terms *flesh*. Flesh is "the layer of tissue sandwiched between the skin and bones" and belongs not to the motor system, but to the sensory system. Flesh is the primary sense organ of touch and acts as "an instrument of tactile and gustatory sensation."[40]

In the centuries that followed Aristotle's writings, picking apart his notion of the *neura* was a major conceptual and anatomical endeavor. The first anatomical structure to be separated out of the bundle of structures Aristotle had designated as *neura* — the rather fibrous tissue that he dissected (hence the translation of *neura* as "sinews") — was a distinct tissue that was to become the nerve in the modern sense.[41] Still named "nerves," but now referring to a more specific anatomical structure, they also acquired a new function. The Hellenistic physician-anatomists Herophilus (ca. 330/320–260 BCE) and Erasistratus (ca. 330/325–255/250 BCE), working in Alexandria in the third century BCE, defined them as the carriers of sensorimotor impulses that traveled to and fro between the periphery and the brain, not the heart. Herophilus and Erasistratus distinguished two kinds of nerves with different functions: they were sensory nerves when sensation was involved and motor nerves (also called voluntary nerves) when motion was involved.[42]

Working in the first century CE and building on the work of Herophilus, the Greek physician Galen of Pergamum (ca. 129–216 CE) described the motor or voluntary nerves as "those structures which transmit voluntary motion from the brain to the muscles."[43] Now motion was transferred via the nerves (in the modern sense) to the muscles, which Galen — "in what was arguably the single most momentous definition in the history of Western medicine" — redefined as the organs of voluntary motion.[44] Galen's "muscles" were the anatomical structure that Aristotle had previously called "flesh," but whereas for Aristotle the flesh had played no part in the organization of locomotion, the muscle now took on the leading role. The muscles were the soul's instrument, and the soul was located in the brain, which Galen identified as the anatomical origin of the nerves and thus as the "source of sensation and voluntary motion."[45] Because the nerves act as transmitters, writes Galen, "none of the parts of the animal has either voluntary motion or sensation without a nerve," and "cutting the nerve straight away renders that part both senseless and motionless."[46] The sole function left for the newly defined muscles was that of contraction and relaxation.

For both Aristotle and Galen, motion manifested the action of an embodied soul, but over the course of half a millennium, all the components of motion had changed. The sinews, once simultaneously transmitters and contractors, were now differentiated into the nerves and muscles, separate anatomical structures with distinct functions, and the soul, too, took on a new shape. For Aristotle, the soul was one, resided in the heart, and inflicted movement on an animal moved by "mind and desire." For Galen, following Plato, the soul was tripartite. Its rational part, designated as the controlling or ruling element of the soul, was responsible for cognition, sensation, and voluntary motion and was located in the brain. Emotions were attributed to the appetitive soul in the heart, and the liver was the seat of the nutritive soul, responsible for nutrition or reproduction.

As Shigehisa Kuriyama has pointed out, Galen's momentous definition of muscles as the organs of voluntary motion made them instruments of the soul, able to express "psychic presence . . . solely through their tensing."[47] In that definition, it was the soul that created the muscle as its own instrument, yet the very same definition meant that the muscle ultimately killed the soul. For once the muscles were anatomically and functionally separated from the nerves, they could also be investigated independently—allowing the soul to be pushed out of the body and its movements.

The Soul in Action

In the seventeenth century, both astronomers and anatomists drew motion into their own purview using lenses. They began to investigate motion with new eyes, and a whole series of treatises now returned to the topic of animal motion. These studies were long considered purely a continuation of mechanism, the founding texts of emerging biomechanics. In them, however, function and teleology, the soul and the machine, are not set at odds, but play in unison.[48]

In Fabricius ab Aquapendente's 1625 *De motu locali animalium*, William Harvey's unpublished notes on the same topic of 1627, and Giovanni Borelli's *De motu animalium*, published posthumously in 1680 and 1681, all three scientists declare from the outset that it is the soul that moves the body. Harvey (1578–1657) agrees with Aristotle that "all movement is derived from the soul" and is differentiated according to the soul's three powers: "some movements derive from the nutritive or natural parts and are common to all living things, others from the animal parts and are proper to animals, others from the rational parts and are proper to man."[49] Harvey also follows Aristotle in defining motion as change in quality, change in quantity, and change in spatial location, driven by the avoidance of pain and the quest for pleasure. He comments that motion is essential for protection and self-preservation, the search for food, and the rearing of offspring.[50] Motion alone is the difference between "what lives and what is dead, between the animate and the inanimate, between a live man and a corpse."[51] In the introduction to his *De motu animalium*, Borelli (1608–1679), too, insists that "the obvious must not be neglected: everybody agrees that the principle and the effective cause of movement of animals is the soul. The animals live through their soul and keep moving as long as they live. When dead, i.e. when the soul stops working, the animal machine remains inert and immobile." Similarly, "it is known that muscle by itself is a dead and inert machine in the absence of an external motive faculty. The latter orders, stimulates the muscle from its lethargy and forces it to move."[52]

It is by the muscles, writes Fabricius, that the soul implements motion in the body. Muscles and joints are the "principal and proximate instruments" of motion.[53] For Harvey, they are what makes it possible to perform movements, and anatomy's remit must therefore be "to define how muscles are moved, how the limbs are moved by many muscles acting in harmony and rhythm, and how the body

44

is moved by the limbs in walking."[54] Fabricius (1533/37–1619) and Harvey set out what Peter Distelzweig calls their "Galeno-Aristotelian" approach at length.[55] They study muscles first according to their structure or fabric (*historia*). Characteristic of Fabricius especially, and in line with Aristotle, are his discussion of the variations of muscles in form, number, and size within and across different species and the teleological explanations that he offers for the presence or absence of particular anatomical structures in one species as opposed to another.[56] Following the distinction introduced by Galen, Fabricius and Harvey then study the action that muscles perform (*actio*) and finally the muscles' use, that is, how they contribute to the life of the animal (*usus/utilitas*).[57] As for how muscles actually work, both agree that their action is contraction and their utility is leverage.[58]

Borelli—a mathematician and mentor of his famous compatriot, the microanatomist Marcello Malpighi—begins his treatise at precisely this point, the mechanical operations of the muscles. He considers not only running and jumping, flying and swimming, but also skating on ice and moving underwater by means of other "mechanical artefacts," such as diving suits and submarines.[59] Borelli wraps the body in measuring tapes, loads it with weights, and calculates its resistance with rods (his plates teem with them); he boils muscles and carries out vivisections, lays dried and living material under the microscope, draws comparisons with existing scholarship, stretches ropes over nails and reels to reconstruct the operations of joints and muscles.[60] By combining the dimensions, position, leverage point, and alignment of muscles and muscle fibers, tendons, and bones, and analyzing them by analogy with "levers, pulleys, winding-drums,"[61] Borelli discovered that bones and muscles work as a lever when a weight is lifted, the muscle representing the lever's shorter arm.

For the first time, Borelli was able to demonstrate that the muscles are capable of actually moving the body with the forces they exert. He devotes much space to the precise calculation of the force exerted by various muscles and required for particular actions to be performed. Among other things, he calculates the force needed by the arm muscles to carry weights and by the leg muscles for simple standing on two legs or on one, for dynamic walking on two or four legs, or for jumping (fig. 1.1). Contrary to his contemporaries' assumption that nature is economical, Borelli concludes that "light weights are carried by large and strong force rather than heavy weights being supported by small force."[62]

Borelli extends the scope of his investigations from external locomotion to the internal movements in animals, because all the vital operations in the organism are "either movements or actions which require movements."[63] In the second part of De motu animalium, he turns his attention to nutrition, digestion, and breathing, along with the work of glands, convulsions, and fevers.[64] But the first movement he tackles is the beating of the heart.

Following William Harvey, Borelli regards the heart as one of the animal's main muscles. Being a "muscle of the same nature as the other muscles," its task is that of contraction.[65] According to Borelli, contraction is caused by swelling, which in turn is the result of fermentation, or more specifically the "fermentative ebullition of the juice of tartar of the blood provoked by mixing with spiritous juice instilled from the nerves."[66] Unlike the muscles of the limbs, however, the heart is not subject to the will; it appears instead to be driven by "blind necessity."[67] In fact, this is why Galen did not consider the heart to be a muscle, since a muscle's movements are voluntary, whereas those of the heart are natural or spontaneous.

Borelli proposes the following solution to this problem. True, he says, in general, the nerves transport the commands of the will from the brain to the muscle, but in the case of the heart, "the nerves are shaken in the brain by another cause different from the will, spontaneously without any rousing." Without further explanation, he continues: "Nervous juice is instilled into the heart by itself. And, therefore, pulsations occur."[68] Borelli attributes the regular intervals of the heart's beating to a special mechanism whereby

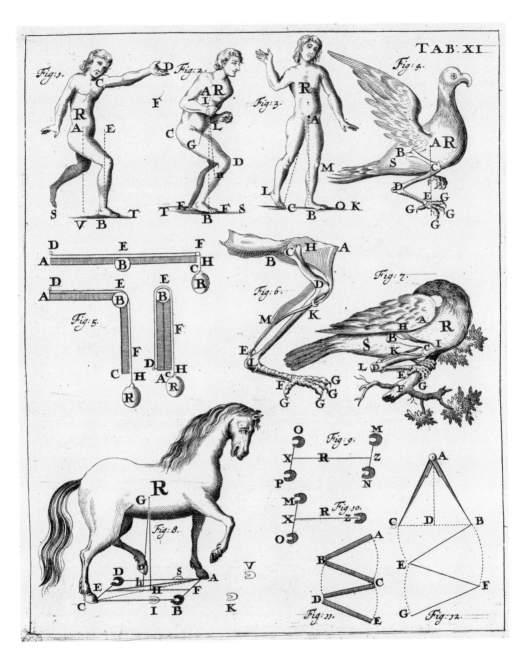

Figure 1.1. Human beings, birds, and quadrupeds standing on one leg and walking. Borelli discusses the differences between humans and animals and the forces exerted in their movement. Borelli, *De motu animalium*, 1680–81.

the nervous juice "is expressed drop by drop into the flesh of the heart."[69] Whenever the droplets have accumulated to a certain volume, the heart will beat in a recurring cycle, which is why the motion of the heart is at once uninterrupted and pulsating.

"This discussion," Borelli first concludes, "sufficiently explains that the movement of the heart can occur by a natural instinct or by organic necessity in the same way as an automaton moves." But if automatic motion can be explained in this way, the heart remains merely a muscle. Borelli therefore uses the artful rhetoric of multiple negation to formulate an alternative hypothesis: "It is not superfluous to consider whether there are some reasons to question if movement of the heart can result not from pure mechanical necessity but from the same faculty of the soul by which all the other muscles are moved." The possibility thus exists that the movement of the heart, too, could be a voluntary movement — one that has become habitual and therefore gives the impression of being natural: "the movement of the heart may be caused by the animal cognitive faculty but unconsciously as a result of acquired habit."[70]

Treating of the soul alone as the obvious principle and cause of movement in the preface to De motu animalium, Borelli dismisses it rhetorically from his further investigation. At this point in the treatise, he now also dismisses the soul de facto from the body. Although Borelli's carefully ambiguous formulations leave open the theoretical possibility that the soul might be the motor even of involuntary movements, his thought experiment strikes at the division between natural and voluntary movement established since Galen's day and hence at the centrality of the soul as the embodied instigator of movement by and in the body. Borelli proposes additional examples — the involuntary movements of the feet or the eyelids to keep one's balance or ward off danger — that underline the doubtfulness of any clear distinction between voluntary and natural. Outside the body, too, no such difference can be identified, since not only does a turtle's excised heart "pulse in a dish for several hours," but the muscles of voluntary movements, too, "move for a long time without receiving orders from the will through the nerves from the brain which has been separated."[71]

Borelli here proposes something conceptually that Jan Swammerdam (1637–1680) tries to demonstrate with experimental rigor. At around the same time as Borelli, in the 1670s, Swammerdam carried out elaborate dissections and important experiments with frog muscles in the Netherlands. This highly skilled microanatomist, working with the finest instruments and sharpest knives, showed that muscles, when contracted, are not swollen as Borelli suggested. "Relying on the propriety and certainty of the experiments I have proposed," Swammerdam reports that he can now, "without any difficulty, maintain that a muscle, at the time of its contraction, undergoes no inflation or tumefaction, from the afflux or effervescence of the supposed animal spirits."[72] By isolating the nerves with the muscles attached to them, he also shows that the frog muscle can be moved by a purely external stimulation of the nerves. This offers experimental confirmation of his earlier observation that muscles isolated from the body can still move. At the same time, Swammerdam rejoices in the opportunities his finding offers for anatomists to "make the muscle contract itself as often as we please." Exactly how the effect on the nerve "is produced from without, and by art," Swammerdam confesses, is a very difficult question to answer, "and perhaps impossible, till the true contraction of the muscle shall be exactly known."[73]

In the mid-eighteenth century, the famous Swiss anatomist and physiologist Albrecht von Haller (1708–1777) proposed a new way of understanding the "true" contraction of the muscle. Haller relocates the cause of movement into the muscle tissue itself. Based entirely on systematic experimentation (painful vivisections, mechanical or chemical manipulation, and the irritation of isolated tissues or organs), Haller's distinction between the "irritable" muscle fiber and the "sensible" nerve fiber created a new terminological, conceptual, and experimental framework that would prove crucial to later nineteenth-century developments in physiology.[74]

Haller held that the movement of a muscle — that is, its capacity for contraction — is made possible by the irritability of muscle fiber. Contraction is a property that Haller determined experimentally and believed to be restricted to muscles alone, but still thought of as purely mechanical. The muscles' mechanical movement, though, is triggered by stimulation of the sensible nerve fibers. Whether a movement is voluntary or spontaneous, Haller concluded that stimulus of the nerves is always necessary.

Haller's work was momentous for investigations of movement because it introduced the physiology of stimulus and response into the study of vital motion. Yet even Haller does not completely jettison the soul. He still attributes to the soul the function of transmitting its will, via the nerves, to the muscles to be moved. However, the soul is now no longer the cause of motion in the sense of muscular movement: Haller regards muscular contractility as purely an intrinsic property of tissue. The soul acts exclusively through the nerves; it no longer steers the dynamic motor organization of the body directly, but acts indirectly, by stimulating the nerves.

Haller closed his lecture series *De partibus corporis humani sensibilibus et irritabilibus*, held before the Göttingen Academy of Sciences, with the statement that "neither Irritability depends upon the soul, nor is the soul what we call Irritability in the body." The soul no longer has any influence on the physiology of the body (now regarded as autonomous) and is limited to consciousness alone.[75] Haller's categorical tone here implies that he considers his work to have drawn a line under previous discussions on the cause of vital motion.

The distinction between irritable and sensible tissue made the contested distinction between voluntary and spontaneous movement obsolete, all muscles, for Haller, being inherently contractile. It also threw the sensitive soul, the last and greatest challenge for understanding motion, out of the organism. From now on, motion and sensation were separated, and the study of *all* physiological movement in and of the organism was a matter of experimentally determining the nature of the tissues involved. Specifically, Haller showed that the vital tissues of the body, especially the heart, are actually more irritable than the tissues subject to the commands of the soul.[76] Now regarded as separate body systems, motion and sensation would ultimately become the objects of separate experimental regimes of inquiry.

That did not mean the end of the soul, which continued to find new forms as a purely immaterial principle of motion and life — in the animism of Georg Ernst Stahl, for example — or conversely, as pure materiality in Julien Offray de La Mettrie's provocative *L'homme machine*, as well as a long and varied series of other "vital faculties," "forces," or "principles."[77] What it did mean was the beginning of a new biology, stepping into the uncharted territory of irritable and contractile matter. Self-motion had been the unfathomed Other of both mechanism and anatomy; now the path was open for experimental physiology to make it the foundation of the body's whole existence.

CHAPTER TWO

Visual Experimentation

In a few brisk pages opening his *Du mouvement dans les fonctions de la vie*, a collection of lectures delivered at the Collège de France in 1868, Étienne-Jules Marey (1830–1904) gives a terse, yet illuminating account of how natural history became science: zoology through Georges Cuvier's comparative anatomy, botany through Antoine-Laurent de Jussieu and his classification of plants, anatomy through Xavier Bichat's histology, embryology through the microscope and the work of Karl Ernst von Baer. As for physiology, it evolved into a science through experiment — specifically, the work of Albrecht von Haller, "who assembled the materials of physiology, made of them a well-defined science, and steered them onto the path of experimentation."[1] Marey, a pioneer of French experimental physiology and photographic technology, thus defines the science of physiology first and foremost through the "How?" of the experimental method. But what is the "What?" of physiological research? What exactly is physiology — or "more correctly," Marey insists, "*biology*, the study of the phenomena that occur in living beings?"[2]

The Science of Living Movements

Toward the end of the eighteenth century, physiology began to see the body as something in constant flux. Carl Friedrich Kielmeyer's (1765–1844) influential *Ideen zu einer allgemeinen Geschichte und Theorie der Entwicklungserscheinungen der Organisationen* (Ideas for a general history and theory of the developmental phenomena of organizations) of 1793–94 describes organic material, "to the extent that it can really be observed," as being "liquid." It is "unformed in itself and mobile in itself; as a whole it is, precisely due to this fluidity, receptive to all forms and all movement and all life in as much as all life consists in movement."[3] How can the organism, in a state of such instability, nevertheless be an organized and functioning whole?

The question of organization, among the most disturbing and intellectually challenging of biology's early days,[4] meant that interest in the organism focused on how its parts worked together. The fact that these parts themselves were highly unstable was still a marginal concern.[5] When Marey named physiology a paradigmatic experimental science in the nineteenth century, he guided attention away from organization back to the instability at its roots and hence to motion as the crucial physiological feature of the organism. "Of all the phenomena that characterize life," Marey writes, "movements are the most important ones; one might even say that in general, movements are what characterize all functions; that it is in this form that the phenomena occurring in animals may be analyzed today with admirable precision."[6] All vital phenomena and all their most important dimensions, in other words, are movements. Motion is the foundation of every process in the body's interior. As such, it is the body's constitutive principle. Experimental physiology is the science of living movements, and its task is to unravel the secret of motion.[7]

Marey lists three defining elements of movements inside the body that physiology must analyze: "their *duration*, their *extent*, and their *force*."[8] A fourth dimension of movement is "its *form*," meaning we require a way of examining motion "that takes account of the different phases of the movement — no longer just its beginning and end, its maximum and minimum — and that determines all its intermediary states." Duration, extent, force, and form are all qualities enabling motion to be described, but only "provided that the movement leaves a trace.'" The form of motion is that constant trace, the trajectory across all the intermediate stages from start to finish, spanning the maximum and minimum. It is only this continuing course that enables us to determine the movement's duration, extent, and force.[10]

To register the tracks, Marey invented all sorts of graphic inscription devices, each dedicated to a particular physiological phenomenon, that "charged the moving body itself with the task of tracing the form of its movement."[11] The first of these instruments was the sphygmograph of 1860, which recorded the pulse, detecting it with a mechanical lever at the wrist, rather than feeling it with the fingertips, and continuously inscribing it as an undulating line on smoked paper.[12] As Marey writes, "devices of inscription" like this one "penetrate the intimate function of organs where life seems to translate itself into incessant mobility."[13]

A Phenomenon Produced by Its Recording Devices

It is something of a commonplace among historians of science to say that the science of life in the nineteenth century was fueled and propelled to new heights by the emerging ideal of mechanical objectivity. A dizzying abundance of automatic styli, rotating smoke-blackened cylinders, cameras, and countless other inscription devices were invented to record the phenomena of nature directly, bypassing the senses and interpretations of the researcher and the prison of his subjectivity.[14] It became obvious that the senses are fallible, unable to paint a universally valid likeness of the perceptible world as it assumedly is. Promptly brought to bear on the mind itself, the new experiments and instruments of sensory physiology and experimental psychology supplied the first empirical proof that perception differs dramatically between different test subjects.[15] Even more problematic than these divergences was the restricted reach of the senses. For physiology, writes Marey, analyzing life experimentally means "reducing to its simplest elements a phenomenon that is too complex to be comprehensible." In the resulting "battle of details," the senses fall short not so much because they vary by individual as because they are incapable of capturing those very details: our senses miss "objects that are too small or too large, too close or too distant, movements that are too slow or too fast."[16]

In Marey's physiology, as the science of living movements, the failure of the senses is in fact more radical. He considers the perception of motion to be utterly beyond the scope of sensory perception. Movements elude the senses not only in degree (being too slow, too fast, too ephemeral), but constitutively: "movement, by its very nature, entirely eludes vision."[17] The study of movement is a science of the imperceptible. Unlike tissues, organs, or bacteria, movements correspond to no object other than one that is generated by technical devices and visual traces; the technological object produces the phenomenon in the process of representing it. Accordingly, Marey's "inscription devices of movements" most often express "phenomena that direct observation has never been able to grasp." At times, they are "designed to replace the observer," but "they also have their own domain where nothing can replace them. When the eye ceases to see, the ear to hear, the fingertips to feel, or else when our senses supply deceptive appearances, these devices are like new senses, with an astounding degree of precision."[18] For the first time, Marey's motion research trained the spotlight on perception, the role of the senses and of the observer, and the representability of movement.

From the perspective of the history of motion, the question posed by Marey's new approach is how the image as such — or rather, the new representability of motion — affected the concept of vital motion in the late nineteenth century. If physiology was the science of motion, and technology could now at last make motion accessible to the eye, what did science do with these images? Or put the other way around, what could be grasped about motion once it became visible? What criteria of visibility could be applied to something invisible? And what did it mean that motion no longer stood for itself, but was considered in relation to a sensory apparatus that, paradoxically, was not equipped to see it?

Instantaneous Worlds

Marey's physiology and its pictorial universe formed part of a surge of visual experimentation that reconfigured a whole culture of attention, vision, and the viewer from the early nineteenth century onward, in laboratories and movie theaters, on city streets and behind the academy's doors, in scholarship and in painting.[19] The 1830s saw improvements in photographic processes that enabled pictures to be developed ever faster and reproduced in ever greater numbers. Just like other goods and knowledge at the time, the newly mass-produced images traveled fast and far, disseminated by telegraph and steam engine, in periodicals and the increasingly quantitative natural sciences. An unlimited wealth of innovations could now circulate at speed, apparently released from the fetters of place and time.[20]

The new opportunities for giving astonishingly immediate visual form to motion, to events that had previously escaped "the human eye's experience,"[21] entranced the nineteenth-century public and left deep marks on culture. When short-exposure photography became technically viable at midcentury, the fashion for "instantaneous" pictures spread rapidly, a source of wonder, admiration, and entertainment for an audience "hunger[ing] for photographic images of action frozen in time."[22] For photography not only wrote with light; it also played with the beholder. Optical toys creating illusions of motion began to appear in every imaginable technical format in the 1830s. Drawings or photos rotated before the viewer's eyes in phenakistiscopes and zoetropes, sped past the fingers in flip-books, or animated brightly lit scenes against the darkness of magic lantern performances — the technical expression of a pleasure in the representation of settings, actions, and vitality itself that had always been at home in the human imagination.[23]

The iconic serial photographs capturing animals, people, and birds in motion that were made by Eadweard Muybridge beginning in the 1870s and Marey beginning in the 1880s partook in that trend. Muybridge, born in England, was an outstanding photographer of the American landscape and had an artist's gaze. His chronophotographs of a galloping horse, made with multiple cameras in the 1870s for Governor Leland Stanford in California and published in 1878, grew from a tradition that embraced the entire cultural history of the horse and its iconography, if not the entire history of art in its struggle with the visual representation of time and motion.[24] At the same time, they helped to build revolutionary new pictorial and imaginary spaces.[25]

As research on his surviving negatives, slides, and notebooks has shown, Muybridge followed the practice of his peers in extensively revising and retouching his negatives. For Muybridge, Phillip Prodger argues, the point of creating the picture "was not that the camera had captured the image, but that the eye could verify it."[26] In their artistic context, the key aspect of the serial photographs was their surface — the visual *mise-en-scène* of a fleeting moment of motion and thus the appeal to the eye. On his 1881 tour of Europe, Muybridge gave well-attended lectures, widely reported in the press, that emphasized the contribution of his photographic technique to the portrayal of motion in painting.[27]

Marey met Muybridge during the 1881 tour. Inspired, he turned to serial photography, and in 1882 became the first to record photographically the flight of a bird. The photographic gun he used was a refinement of an invention by the astronomer Jules Janssen, who had charted the transit of Venus on a moving photographic plate in 1874.[28] In Marey's hands, chronophotography was to become a data tool.

Marey's Moving Data

Describing the advantages of photography at many points in his work, Marey does not ignore the technique's representational

aspects. Chronophotography is "a faithful memory,"[29] which gives
"an instantaneous picture of the most diverse objects" and re-cre-
ates "the appearance of natural objects," albeit "as seen by looking
with one eye only."[30] Photography in general, he notes, is increasingly coming to "replace outline drawings."[31] Most importantly, the
method substantially expands the potential domain of investigation,
because it "demands no material link" — unlike the other inscription devices Marey invented, photography can be utilized even
"when the moving body is inaccessible, when it cannot be fastened
by mechanical means to the recording apparatus."[32]

Despite these special virtues of photographic representation,
Marey saw chronophotography first and foremost as the logical
continuation of the "graphic method" he had developed over so
many years.[33] Although he does call his photographs "images" or
"pictures," more often he describes them as "curves," "records" or
"written records," "tracings," and "photographic diagrams" or "photographic registration."[34] All the various forms of representation
Marey proposes share a single purpose: to convey information.[35]
Lines, graphs, machines, photographs — Marey regards them all as
vehicles for obtaining data, as points of access to particular phenomena, and as instruments of a science that could not exist without them and their production of visibility.

As Marey himself knew very well, the plotting of curves had
been possible since Descartes, and with it the geometrical definition of time-space relationships or more generally of the relationship between two magnitudes. In *La méthode graphique*, Marey
gives a detailed review of how graphs can be deployed to obtain
information on everything from population, to trains, to epidem-
ics. Unlike such curves, chronophotography does not record the
moving body directly, but projects its course onto "the surface of a
sensitized plate."[36] Whatever the technical form, however, the pre-
eminent requirement that Marey places upon the visual registration
of motion is precision. The chosen technique must both register
the movement's trajectory and "define the various positions of [the]

body on the trajectory at any particular moment."[37] To identify the movement with precision, in other words, it must be possible to determine the relationship "between the distance traversed and the time occupied" at every point in time.[38] For this reason, chronophotography will really become a "method" only if "the two notions of time and space can be combined in photographic images." In that case, we will "have instituted a chronophotographic method, which explains all the factors in a movement which we want to understand."[39]

From Physics to Perception

In Marey's human and animal locomotion studies, the space-time relationships of a movement cannot be fully specified if each image is made separately. Aiming to reveal the successive phases of the bird's flight or the human gait in as many segments as possible and at identical time intervals, Marey finally decides to photograph all the movement's stages on a single plate.[40] These images, which he sometimes calls "partial photographs," use a new procedure: the subject's body is made to disappear by being dressed in black against a black background with just a few "bright stripes and spots" applied to particular limbs, the only items to be registered by the camera (figs. 2.1 and 2.2).[41]

Here, the analysis of the movement consists in identifying the relationships between individual body parts, such as the shoulder, knee, or thigh. In this arrangement, Marey's traces, records, and diagrams can fulfill their true purpose, measurement. They may measure the motive force of the leg muscle during walking; the duration, phases, and intensities of the horse's hoof hitting the ground or taking off again; or forces such as heat, mechanical work, and electricity.[42] Marey regards all the various visualization techniques as steps on his path to making motion a means of measuring physical forces.

Firmly antivitalist in outlook, Marey sets out to demonstrate that the organism is subject to exactly the same physical forces

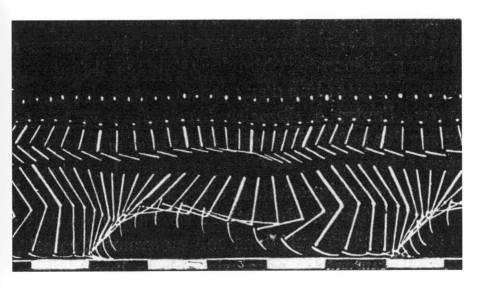

Figure 2.1. A "geometrical chronophotography" of a man dressed in black velvet with reflecting stripes attached to his limbs and photographed against a dark background. Marey, *Le mouvement*, 1894.

Figure 2.2. White paper is fixed to the tips of a crow's wing. The bird flies in front of a dark background while the camera captures the light reflected from the paper. The arrows indicate the direction of flight. Marey, *Le mouvement*, 1894.

and laws as those governing every other form of matter. The body is nothing but a movement machine, and Marey's aim is to evaluate its performance. This was an interest shared by many others in the early twentieth century, whether the military, the German gymnastics movement, or the emerging science of ergonomics. Marey's work paved the way for new methods, cinematographic analyses, and studies of hand movements, work flows, and physical efficiency that would be applied to industrial production by Marey himself (working with the engineer Charles Fremont and ergonomist Jules Amar), Nikolai Bernstein at the biomechanics laboratory of the Central Institute of Labor in Moscow, and Frank Bunker Gilbreth in early twentieth-century US manufacturing and engineering studies.[43]

Marey's locomotion studies have often been interpreted, and rightly so, as a hinge point between natural motion and rationalization, leisure and labor, creativity and productivity—between body and machine, aesthetics and biomechanics—in the upheavals of modernity. Discussions of his work have linked it with the Weber brothers' early *Mechanik der menschlichen Gehwerkzeuge* (*Mechanics of the Human Walking Apparatus*, 1836) and a late nineteenth-century collaboration between the Leipzig anatomist Christian Braune and the mathematician Otto Fischer, *Der Gang des Menschen* (The human gait, 1895).[44] In France, Marey's research evolved within a historical and literary tradition that embedded everyday bodily movements into a multifaceted discourse on everything from Rousseau's experience of nature and the cultural history of travel to gestures, expressive movements, and normal or pathological physiologies.[45]

Human and animal locomotion was just one form of motion, and running soldiers, high-jumping athletes, and flying birds were easy to see. Yet Marey had defined *all* the vital processes in the organism as motion and made physiology—as a general science of living movements—dependent on experimental methods and visualization. If we turn to this broad definition of physiological motion, to imperceptible movements inside the body, it is the epistemological dilemmas facing him that come to the fore. Marey had to combine the uncombinable in multiple senses: his methodology of measuring motion with its presentation to the eye; his habitus as a physicist with his experimentation on living bodies; his firm belief in the shortcomings of the human senses with his lifelong dedication to a phenomenon posing one of the greatest challenges to human perception. These dissonances draw us away from biomechanics to a very different domain of knowledge — psychology, and specifically the psychology of sensory perception.

The Eye and the Imagination

The perception of motion is beyond the reach of sensory perception alone. Because of the deficits of our own sensory apparatus, the tracings, records, and images of physiological experiments offer us our only access to data, our only means to define them. But although Marey (as he frequently stresses) is interested in the image purely as a supplier of data, his experimental method gives rise to pictorial forms that exist as images and affect the viewer as such. Qua image, they unfold their own epistemic dynamics, operate beyond the experimental framework, inscribe themselves into traditions, and open up new viewpoints. Marey's awareness of this pictorial dimension of his work is indicated by his disdain for cinematography as a means of synthesizing motion. His survey of methods of motion research in *Le mouvement* of 1894 (*Movement*, 1895) culminates in the question of how a movement, once dissected by chronophotography, can be resynthesized in the moving image. If the aim is to reproduce natural movements that can anyway be tracked by the naked eye, an appeal to the sense of vision is "simply childish," being still beset "by all the uncertainties and difficulties that embarrass the observation of the actual movement."[46] Writing in the very early days of cinematography, he argues that unlike actual moving images, chronophotographs rely on "the imagination," rather than "the senses."[47] At the heart of knowledge of motion, for Marey, is not the appeal to the eye — the aspiration of an actually moving image — but the appeal to the imagination.

In Marey's inscriptions, we see a movement taking on form, and as it does so, we become conscious of that very formation as an act of imagination. The process by which motion becomes form calls not for vision as a gaze, but for vision as an act of perception. Given the impotence of our own senses as an instrument, the interaction of vision and imagination in perceiving movement has to be mediated by the devices of inscription. Erin Manning regards nineteenth-century images of movement as attempts to investigate perception through an apparatus and simultaneously as perceptual experiments in their own right.[48]

This ambidexterity explains the fascination that the images still exert today, for in Marey's photographs, "the elasticity of perception is felt."[49] Marey's visualizations therefore depart from the home ground of visuality. They add to vision another force, that of perceptual synthesis. Manning writes: "We see holes: contours, edges, active intervals moving. We feel wholes: experiential duration not yet divided into actual objects."[50] The imagination draws viewers of Marey's images into an active encounter as the capacities of picture and perception oscillate between what is palpably in the offing and what is not shown — or at least not yet. Art has always sought to capture that moment, but Marey's work marks a turning point where it is no longer negotiated between the viewer and the represented alone.[51] A third element joins them, the machine.[52] In the seeing of motion and in motion's taking form, there is a mutually constitutive relationship between human being, machine, and representation.

In the nineteenth century, technology helped to overcome the existing limitations of representability. At the same time, research into sensory perception in general — and vision in particular — moved from optics to physiology and thus from the exterior into the interior, into the "thickness" of the body.[53] In this sense, Marey's representations carry out research at the margins of the body — at the threshold between self and other, between the physiology of perception and the physics of motion. These frontiers were able to shift because the spectator's body was contemplated as a perceptual apparatus on its own account, one in which the functioning of the human eye is no different from that of the light-sensitive plate. The photograph acts as the "retina of the scientist": an interplay of light and inscription akin to the interactions of nerves and stimuli that were studied in experimental physiology and psychology.[54]

Once motion was tied to perception and thereby pushed inside the body, the scene was set for the new field of psychological motion research to arise—but not until a century had passed. The intellectual and aesthetic puzzle of holes and wholes would gain new momentum only with the advent of a new machine, the computer. In the second half of the twentieth century, the science of the computer started to change the science of life profoundly. It was in 1973 that the term "biological motion" appeared. Ironically, the new term to capture the specificity of organic motion had arrived just as that very specificity was about to become eroded.

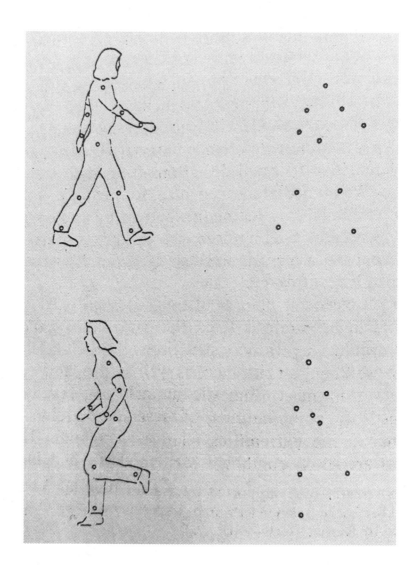

Figure 3.1. A walking or running person with reflective patches attached around their joints is recorded on video (*left*). On the screen, only individual bright spots are visible (*right*). Observers begin to interpret the dots as a walking or running person as soon as the images begin to move. Johansson, "Visual Perception of Biological Motion and a Model for Its Analysis," 1973.

CHAPTER THREE

Calculating Perception

I have talked about motion as something that happened to all bodies in the new science of the seventeenth century and as the defining feature of the science of living bodies in nineteenth-century physiology. In this chapter, I turn to the twentieth-century fascination with living bodies *processing* motion and ascribing it to other living bodies. In 1973 and 1976, two essays by the Swedish psychologist Gunnar Johansson (1911–1998) "completely hooked" his colleagues and laid the foundations for a new way of scientifically studying animate motion.[1]

Perceptual Wholes and the Psychology of Events
Johansson was the first to notice people's ability to identify human motion on the basis of extremely reduced stimuli. Although he never mentions Étienne-Jules Marey in his work, Johansson's experimental methodology immediately brings Marey's to mind, and at first sight, his studies appear to be a seamless continuation of Marey's motion analysis. Like Marey, he used point-light displays to eliminate as much information as possible about the moving object and its motions (fig. 3.1).

Johansson applied bright spots to the joints—initially light bulbs, later reflective adhesive tape—and filmed the resulting traces of light against a dark background. He demonstrated that dramatically reducing the number of stimuli (the points of light) did not impair perception of the motion. Far from it: Johansson's test subjects were able not only to identify the illuminated traces with almost complete certitude as human movements, but importantly, also to distinguish particular types of human movement. The observers in his experiments integrated different information derived from the individual points with an impressive degree of accuracy, indeed almost flawlessly, to perceive a human being in motion — more precisely, a walking, running, or jumping human being, and more precisely still, a melancholy or cheerful, male or female human being.[2]

By using degraded stimuli, Johansson showed that the visual system is able to transform isolated, apparently disconnected moving points distributed in space into coherent percepts. In his 1973 article "Visual Perception of Biological Motion and a Model for Its Analysis," Johansson presented "the first phase of a research program on visual perception of motion patterns characteristic of living organisms in locomotion" and explained that "such motion patterns in animals and men are termed here as biological motion."[3] His interest in biological motion revolves around the question of how the observer integrates the individual points of motion into the perception of a moving unity. That concern was not new; he had already introduced it in his 1950 doctoral dissertation, published as *Configurations in Event Perception*. This early work proposed events as an object of psychological research, using the term "event perception" for the first time in an English-language study.[4] Describing his experience of watching a windblown birch outside his window, Johansson defined an event as "the interplay of continuous changes ... whose carriers can be perceived as spatially separated, occurring simultaneously in the same perceptual field." He asked: "Does the psychology of perception enable us to state why the tree in the photograph, this confusion of points and lines, appears as a closed whole; separated from its surroundings; and to explain the reason for the organization of the internal tree pattern?"[5]

At the core of an event, for Johansson, is motion and the organization of stimuli into a "perceptual whole." In our perception of the birch, "the motion has at the same time had a uniting and a segregating effect."[6] Certainly, Johansson's choice of example typifies his dissertation's interest in motion very generally, not animal motion as such. However, his notion of the perceptual event introduced a new theory of visual perception that not only had nothing in common with Marey's view of the senses and their imperfections, but also "differed fundamentally" from the models current in the 1950s.[7] It was this theory of perception based on motion that would pave the way for his new concept of biological motion some twenty-five years later.

A Mathematical Theory of Perception

At the start of his career, Johansson worked in Stockholm with the Gestalt psychologist David Katz, an exile from Nazi Germany.[8] He was influenced by Gestalt psychology and by Gestalt-related work that considered the perception of the incomplete as a whole and, from that perspective, the phenomena of motion perception and motion illusions. Max Wertheimer, for example, wrote his 1912 professorial dissertation on the illusion known as the phi phenomenon, in which a rapidly alternating stationary representation is perceived as motion; Karl Duncker published his seminal study of induced motion in 1929; and Albert Michotte, in his famous *The Perception of Causality* of 1946, investigated simple interactions evoking the impression in the viewer that one object triggers the motion of another.

Unlike these scholars, Johansson addressed not motion illusions, but actual locomotion. The novelty of Johansson's theory of spatial and motion perception, which radically changed the field,[9] lies in its mathematical approach. It is based on the geometrical procedure of central projection, not Euclidian geometry. In central projection, one plane is projected onto another from a vantage point on neither of the two planes—a perspectival procedure that comes very close to the eye's natural perception of the environment, such that if the vantage point corresponds with that of the human eye, observers can take a bird's eye or a worm's eye view, or anything in between. Within this geometrical framework, Johansson proposes a model of movement perception based on three principles: elements moving on one plane are "always perceptually related to each other"; equal and simultaneous motions "automatically connect these elements to rigid [that is, relatively constant] perceptual units"; and, crucially, elements from which motion vectors can be mathematically abstracted "are perceptually isolated and perceived as one unitary motion."[10] Without going into further detail here, the point to note is that the perception of motion obeys the rules of central projection and the mathematical calculation of motion vectors. In short, "vision functions in accordance with strict mathematical principles."[11]

Johansson's theory of visual vector analysis explains the coherent perception of motion on the basis of variable stimuli, because "the ever-changing stimulus pattern on the retina is analyzed to maximal rigidity in coherent structures."[12] The observer integrates the individual points of motion and sees them as a "perceptual unity," but now in terms of their "relative motion" against a reference frame.[13] Biological motion is thus no longer considered in terms of the spatial shifting of the body's gravitational center, as it was in the mechanical analysis of motion from Borelli in the seventeenth to Braune and Fischer in the late nineteenth century. Rather than with spatial dislocation, motion begins with the deformation undergone by an inherently mobile, unfixed, biological body.

Defining Biological Motion

Gunnar Johansson was a well-connected and influential figure in postwar Western experimental psychology, pursuing both basic research and applications in aviation and traffic. To this day, his work is regarded as foundational for research on biological motion. But why did Johansson turn from the motion of objects to the motion of human beings, from birches to bodies? And what prompted his momentous innovation, the term "biological motion"?

Today, the field has largely moved away from Johansson's original concerns. Research has shifted in the direction of visual social perception, an area that Johansson himself ignored, and biological motion is now regarded as a source of social information. Human and animal movements supply a plethora of socially relevant information that attends all human interaction, and the perception of movement is one of the oldest survival mechanisms in evolutionary terms.[14]

But Johansson was interested in none of this, and current research trends in biological motion do not greatly illuminate his original intention. In fact, little significance has been attached to his choice of the phrase "biological motion" by researchers since then. Nikolaus Troje sees it not as a technical term that Johansson specifically wished to introduce, but as an "informal way to refer to the stimulus domain from which he derived the displays that he needed to further his studies on the role of motion in perceptual organization," the epithet "biological" being primarily chosen to describe "a very interesting stimulus class for the vision researcher."[15] James Cutting argues that Johansson made "no sharp distinction among events, biological motion, and human movement. For him they were all complex 'continuous changes.'" He surmises, though, that biological motion involves organisms perceiving other organisms (whether human or animal) moving, and thus excludes Johansson's weeping birch from the world of biological motion. Interestingly, Cutting makes this argument not, as one might expect, on the grounds that the tree lacks perception, but on the grounds that the motion "is not self-motivated by the tree."[16] Cutting — who knew Johansson personally and is himself noted for his research in the area — here leaves aside perception and appeals instead to the very oldest of the many topoi distinguishing organic motion from any other motion, self-motivation.

What, then, was Johansson's rationale for referring to biological motion? On the one hand, his terminological choice does seem to set biological motion apart from other forms of motion. On the other, Johansson himself certainly regarded the human movements he investigated in the 1970s as supplements to the mechanical examples he had studied earlier in his career, further proof of how the human organism can organize dynamic perceptions. The transition seems quite simply to have offered heightened complexity, allowing him to extend his vector theory from the perception of mechanical movements to the far more sophisticated perception of biological movements.[17] I would suggest a further reason, however, one that Johansson himself mentioned only casually when reviewing his life's work and that emphasizes the "biologicalness" of motion from a different and unexpected perspective: the advent of the computer.

From Registration to Calculation

As we have seen, Johansson's point-light displays initially seem to continue Marey's experimental studies: optically registered data, consisting of very reduced visual stimuli, capture something that is otherwise the preserve of perception alone and not accessible to analysis. In both scholars' work, representation gives rise to measurement. But unlike his nineteenth-century predecessor, Johansson does not seek to understand perception in analogy to the optical technology used to investigate it. Those older theories, Johansson writes, "applied *pictorial* analyses of the optical stimulation and mainly were dealing with analyses of static 'retinal images.' This holds true for the traditional Berkeley-Helmholtzian approach as well as for the Gestalt psychology."[18]

Now, mathematics steps up to take the place of camera, film, and pictorial analysis. Rather than registering a movement, argues Johansson, the eye *calculates* it. It is not the case that the eye follows and constantly registers a movement that becomes a movement only in contradistinction to its static background. Our perception of a movement does not occur in relation to a fixed backdrop. For Johansson, change is the essential characteristic not only of motion, but also of the space in which motion happens; we are constantly observing things that are inherently in motion, and in motion within an equally mobile environment. How, he asks, can a coherent percept emerge under those conditions? To explain motion is "to explain how space can be experienced as rigid and unchanging in spite of continuous changes in excitation."[19] The perception of motion is the outcome of an enormously complex mathematical analysis of motion vectors in space, and the integration of movements in relation to a reference frame is no more than the result of that process.[20] Motion perception thus resembles not a visual trace, but a mathematical operation, and motion is not captured by a camera, but processed by a brain. As Johansson puts it in the early article "Rigidity, Stability, and Motion in Perceptual Space," published in 1958:

> Systematic experiments with motion constellations demonstrate clearly the inadequacy of taking the static space as the single frame of reference for analysis of motion percepts. Experimental results indicate that our visual apparatus does not function like a cinematographic camera, registering motion tracks, but rather in close analogy with a computer capable of calculating and summing differentials and integrals of motion vectors inherent in the "real" motion tracks of the proximal stimulus.[21]

Marey's chronophotography appealed to the viewer's imagination, motion perception being what we see, but equally what we have to imagine happening between the moments captured in the image. In this way, Marey opened motion research to the "elasticity" of perception based on visual representation — what Erin Manning describes as seeing holes, but feeling wholes. Johansson's innovation was to introduce the "plasticity" of perception based on "mathematical descriptions."[22] By thinking motion mathematically, as if processed by a computer, Johansson is able to jettison the image, which for Marey was still inextricable from movement perception. Optical technology, registration devices, and the camera image do not record perception, for perception is "not factual, bound to an image" — it is mathematical, and the mathematics of description is just as multiform and rich as perception itself. As a result, subjects do not necessarily choose the "most elegant" mathematical description, but may just as well derive a description from their "past experience and actual expectancy."[23]

In the 1950s, when Johansson began to work on his mathematical theory of vector analysis, it met with a cool reception, apparently diverging "in a too radical way from the traditional analyses." That changed with Johansson's investigations of biological motion from the 1970s onward. One reason, Johansson came to believe, was the new interest in visual motion perception among computer scientists, which "started something like a creative paradigm shift."[24] Johansson's theory of vector analysis had introduced mathematics into motion perception during the 1950s; the term "biological motion" took shape at a time when computers were beginning to offer opportunities to work out the mathematical descriptions of motion perception that Johansson proposed. To historians of technology, this may seem like just one more example of a typical itinerary: Johansson started studying human motion perception and using the concept of biological motion at the very moment when computers became able to execute his mathematical calculations mechanically and to cope with the extreme complexity of such calculations in the case of biological movements.

However, although the computer performed the mathematical calculations, it was mathematics that opened up a new space of thought. To think perception mathematically was to change ways of thinking motion and animateness dramatically—not because the computer would now be able to carry out the math, but because the math made it possible to conceptualize life differently. Johansson's work marks the first point in the study of organic motion when attention ceased to focus on the contrast between the movement of organisms and the movement of mechanical things. As we have seen, both Johansson himself and present-day researchers regard his work as addressing an unbroken continuum between those two poles. Furthermore, Johansson adopted an entirely new stance on the abiding problem of how to represent an ephemeral phenomenon visually. Marey grappled with visual strategies to reveal that which eludes perception, but Johansson takes the opposite path. For him, it is the plasticity of perception itself that organizes the image — and more than that, the world itself. "Biological" in this view does not refer only to the specific psychophysiological sensory apparatus of a perceiving observer, whether human or animal, or to the specific movements of people or animals. Rather, and importantly for later developments, the concept acknowledges that it is the perceiving observer who performs the ascription of movement to the categories "organic" or "nonorganic." For the first time, motion could now be formulated as a problem not of existing categories to be analyzed, but of ascriptions to be made.

Johansson's designation of motion perception as "not factual" may have been casual, but it is of outstanding significance for twenty-first-century research in biological motion. Developments in the fields of robotics, biotechnology, and medical research show that the crucial question today is: What happens when spontaneous motion ceases to define the animate?[25]

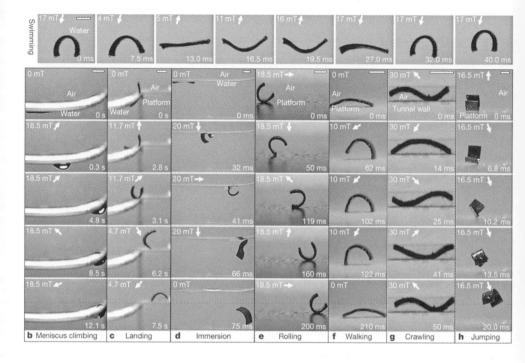

Figure 4.1. Locomotion of the soft robot. Hu et al., "Small-Scale Soft-Bodied Robot with Multi-modal Locomotion," 2018.

CHAPTER FOUR

Silicone Motion

On January 24, 2018, the *New York Times* "Science" website carried a video report on brand-new research at the Max Planck Institute for Intelligent Systems, Stuttgart. In it, the journalist describes a "tiny strip of something rubbery," black, flat, and elongated, a few millimeters long. Thus far, the object is just small and inconspicuous, nothing more than a sliver of matter. But "then it starts moving."[1] The moment it is animated, matter becomes an event. The unprepossessing scrap of rubber can suddenly swim like a jellyfish. It can bend, twist, and roll like a caterpillar, walk like an inchworm, creep through obstacles or leap them like a nematode. The *New York Times* headline comments: "This tiny robot walks, crawls, jumps and swims. But it is not alive."

Soft Robots

This elastic marvel is a microrobot (fig. 4.1), described in the original *Nature* paper as a "magneto-elastic, rectangular-sheet-shaped soft robot . . . made of silicone elastomer."[2] The silicone is embedded with programmable magnetic microparticles, which begin to move when different magnetic fields are activated. The material propels itself, so to speak, but it is guided from outside.

The robot's motion arises entirely from alterations to the shape of the flexible silicone strip. The magnetic field can bend the sheet into a sine or cosine form, for example, creating "a travelling wave along the robot's body." Intensifying the magnetic field switches that sidling into a C or V shape, and alternation between the two states has the robot "enacting a gait similar to jellyfish swimming."[3] This alternation is also what enables the robot to pick up cargo at one location and release it at another. Curved into a C, the robot can climb up a water surface and turn about its own axis. When alternating movements are clustered into "a fast sequence of downward bending, rotation, and flipping," the robot stops climbing and instead dives down into the liquid. In a move inspired by the escape behavior of caterpillars, it can "roll directionally over a rigid substrate" or "dive from a solid onto a liquid surface." The robot's ability "to walk in a desired direction," in turn, is inspired by the inchworm. Walking, defined by the *Nature* researchers as a "particularly robust way to move," is achieved by the "precise tuning of stride length and frequency." It can be modified in the face of obstructions, shifting into the "undulating gait" of a caterpillar in order to "crawl through the obstacle" or jumping over it like a nematode if that proves impossible.[4]

The well-established term "robot" as used in this kind of research evokes a machine. According to the *Oxford English Dictionary (OED)*, a robot is "a machine capable of carrying out a complex series of actions automatically, especially one programmable by a computer"; it may resemble a human being and be "able to replicate certain human movements and functions automatically." The German standard dictionary *Duden* also refers to the human being, defining the *Roboter* as a replacement and surrogate for human activities that may also be "modeled on the human form." Interestingly, the definition of *robot* in the French *Larousse* dictionary as a "machine with a human appearance, capable of moving, carrying out actions, speaking" still gives prominence to the context of science fiction, as does the *OED*.[5]

Ever since the first automata were built in antiquity, the desire to replicate natural locomotion has been motivated by the mechanist notion that copying a natural system within a technical system can help us to understand the natural system more fully.[6] That is certainly true of the currently flourishing research on biomimetic locomotion, which engineers lifelike forms of movement in ingenious technical devices that more or less closely resemble cockroaches, flies, or rattlesnakes in order to understand and consequently make use of the creatures' movements. In soft microrobots such as the one I have just described, however, all formal similarities with the elegant automata of the eighteenth century or the heavy machinery of the nineteenth — as well as with the human or animal body itself — have been obliterated. Sheer matter is the new machine; the material is the technology. The guiding principle of soft robotics is the self-deformation of material. "Robot," here, means matter with the property of activity.[7]

Like more conventional forms of robotics, soft robotics and microrobotics are concerned with the imitation of movement processes. But whereas the classic robot is modeled on an original to copy that original's movements, in the case of our crawling soft robot, it is only by identifying a particular kind of walking, swimming, or creeping that we begin to suspect we may be looking at a jellyfish or a caterpillar; the material is ascribed vitality only on the basis of its motion. Externally controlled and devoid of visual or physical similarity with a living creature, the tiny object is difficult to read. Its movement alone determines what we associate with it — and, consequently, whether we regard it as alive.

The robot, then, plays with our sense of the animate, all the more so because of its movements' versatility. In an object of such unassuming simplicity, nothing more than a flat scrap of rubber, the elegance and variety of its motion is particularly alluring. Ecologically specialized motion behaviors, distributed right across the animal kingdom, seem to have been synthesized in a single tiny piece of silicone. The plain, rectangular outward form permits no inferences or analogies with feet or wings, bipeds or quadrupeds, tentacles or fins — our traditional imaginary of animal motion — and the movement's artistry becomes an outright attack on our taxonomies. What are we looking at? What criteria can we use to identify what we see? Initial bafflement gives rise to the epistemological question: What is motion creating as organic here? For even today, the Aristotelian equation of motion with life governs our ideas of what is alive. The *New York Times* heading "But it is not alive" alludes to those historical roots, yet the research covered in the article seems to imply the very reverse: this tiny robot actually is alive, *because* it walks, crawls, jumps and swims.

On Being Tiny and Moving in a Liquid

To what ends do these miniature robots move with such agility? The hopes attached to them are immense, though still far from realization. The robots' future tasks are to include the target-precise delivery of drugs or genes to the body's organs, tissues, and vessels; minimally invasive surgery; single-cell manipulation; and biosensing.[8] Robots are expected to facilitate medical intervention at sites inside the body that are closed to the human gaze, inaccessible except through technology. In its digestive, urinary, and vascular system, the body is pathless, serpentine, messy — a conjuncture of radically different, continually self-calibrating micromilieus. Once it enters that terrain, the robot is left to its own devices. It must find its way nimbly through the dark, operated by an invisible hand.

Of course, the robot is not the only thing moving inside the body. Everything that happens there is a process of motion. Motion occurs inside the cell, on the short hops between cells, on the long hauls across the body. Cells and subcellular structures move in order to migrate and to transport substances or molecules. New experimental and optical technologies have made it possible to penetrate ever deeper into the subcellular mechanisms with the aim of revealing the dynamics of organic life — specifically, the mechanisms by which cells and cell structures convert chemical energy into mechanical work and motion to perform their respective biological tasks. In eukaryotic cells, the cytoskeleton gives the cell structure and shape, but also enables it to move from place to place.[9] The cytoskeleton is a dynamic mesh of filaments — dynamic in the sense that it can be built up or dismantled — or more precisely, a network of proteins. The movement of and inside eukaryotic cells is produced by microtubules (cylindrical, relatively rigid hollow bodies) and actin filaments, along with the motor proteins dynein and kinesin, bound by the microtubules, and myosin, bound by the actin. Dynein and kinesin are responsible for motion and transportation within the cytosol; the interaction of actin and myosin generates muscular movements.[10]

Motion in the body's interior takes place under difficult conditions. The body is an open system in ceaseless exchange with its surroundings. When something happens inside the body, it always happens simultaneously with hundreds of other biological processes, and that in the tiniest, viscous environment. Furthermore, not everything that moves within the body belongs there — viruses and bacteria swarm. Since not all microorganisms are harmful, and some are harmful at certain locations but harmless at others, the demands placed on the sensors of the cell and subcellular systems and on the motion mechanisms are enormously complex.

Finally, and importantly, microscopic motion is subject to different physical laws from those of macroscopic motion. At the microscopic level, movement occurs at low Reynolds numbers. The Reynolds number, the ratio of inertial to viscous forces as applied in fluid mechanics, determines flow patterns in different fluid environments, such as the flow of air along an aircraft wing or, in our case, the movement of tiny structures in the body's liquid micromilieus. Turbulence occurs at high Reynolds numbers, whereas movement at low Reynolds numbers tends to be a smooth and continuous, laminar or sheetlike flow because of the dominance of viscous forces, as opposed to the inertial forces that govern the movements of large bodies in water or air. Movements in these small-scale environments are extremely difficult to describe in either physical, mathematical, or computational terms.[11] Epistemologically, moreover, they pose a very particular challenge: If, unlike in the macroscopic world, driving forces are absent inside the body — if stillness reigns there — then how do microorganisms move?

Mobile Robots, Motile Cells

In the past fifty years, the issue of motion has inched from biology's periphery to its center, and today, it drives a multiplicity of biotechnological and biomedical inquiries. Examples include research on amoebae in microbiology, on leukocyte motility in immunology, on cell migration during morphogenesis in embryology, on metastasis in cancer research, on motile fibroblasts in wound healing and regeneration research, and, in medicine, on targeted surgery and drug delivery.

Precision medicine is a research field that seeks new ways to enhance the scope and efficacy of traditional therapeutic procedures by targeting medical interventions in the body. Key to this project is the transportation of substances (drugs, nanoparticles, genetic material, stem cells) to the exact site in the body where their action is required. Alongside methods such as molecular targeting, which blocks the growth of particular molecules involved in the formation of cancer cells or cardiovascular disease, researchers are trialing the use of microrobots that can be activated and controlled from outside the body. As well as synthetic soft robots, researchers are now producing biohybrids, combining synthetic components with natural cells or with microorganisms such as bacteria or algae (fig. 4.2). Natural motility is put to work, and research focuses on how, in what form, at what speed, and overcoming what barriers cells or microorganisms can be moved around the body, carrying with them medically relevant agents. The cell itself becomes an animate technical artifact, a "live microrobot."[12]

Here, the boundary between matter and organism has shifted irreversibly, for the microrobot is not a merely "programmable" machine, but a machine "with partly or fully self-contained capabilities for mobility, sensing, and operation of predefined tasks."[13] And

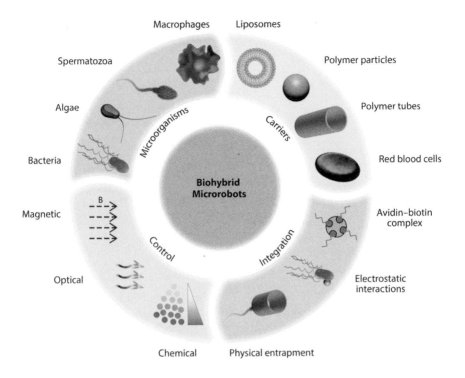

Figure 4.2. The various components and methods of biohybrid construction. Alapan et al., "Microrobotics and Microorganisms: Biohybrid Autonomous Cellular Robots," 2019.

when the body's own forms of motion are harnessed, the epistemological questions raised by synthetic robots become more intricate. It also becomes particularly obvious that the body's interior is not simply in motion: the exact type of its movements and their diversity are crucial to its organization. As we will see, it makes a difference whether cells swim, crawl, swarm, or navigate long distances, whether they are passive travelers or speeding rockets.

Swimming, Crawling, Rolling
The body's immune system repels pathogens invading from the environment—bacteria, viruses, fungi, parasites, or toxins—and

fends off danger from its own defective cells. Cell motion is vital to that system's logistics. A throng of highly specialized immune cells and mechanisms work in concert. Carefully synchronized and sequenced, at different speeds and equipped with different sensors and properties, the cells travel through the body to the locations where their particular services are required.

The most salient feature of the movements of leukocytes (white blood cells) is their adaptation to cross biological barriers between the micromilieus in the body's interior. They can actively migrate into and through tissues and organs — for example, from the blood and vascular system into inflamed or tumorous tissue. The leukocytes include neutrophils, monocytes, macrophages, and lymphocytes. Monocytes live freely in the blood for one to three days before they migrate into tissue, where they will survive for weeks or months as macrophages, surrounding and digesting harmful microorganisms. Most common among the specialized immune cells in mammals are the neutrophils. These have a short life span, between a few hours and a few days, and usually arrive first at the site of an infection, within just a few minutes.[14]

Leukocytes move using a whole range of mechanisms. Their kinetics are "still poorly understood,"[15] but one of the basic types of leukocyte motion is amoeboid movement, a kind of crawling that occurs when the protrusion of cytoplasm forms pseudopodia — literally, false feet.[16] These highly mobile immune cells are adapted "to navigate long distances in the bloodstream" by "passively traveling along" until they reach their required location.[17] There, they transmigrate into the tissue in a sequence known as the leukocyte adhesion cascade (fig. 4.3).

The cascade's various phases — "capture," "rolling," "slow rolling," "arrest," "adhesion," "crawling," "transmigration" — describe in a simplified form how the leukocytes first make contact with the endothelium, then roll along it, gradually slow down and stop, attach themselves to the vessel surface, and finally move through it.[18] The movements of the leukocytes are steered by chemotaxis,

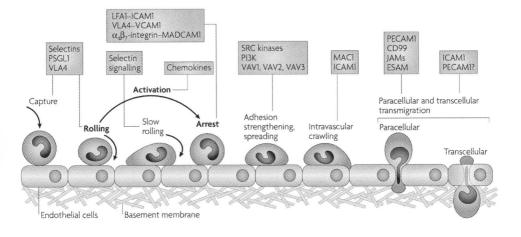

Figure 4.3. The leukocyte adhesion cascade. Ley et al., "Getting to the Site of Inflammation: The Leukocyte Adhesion Cascade Updated," 2007.

that is, their response to gradients in the concentration of chemical substances, and by "contact guidance," reacting to the physical and chemical properties of the substrates on which they move.[19]

One medical application is to have the leukocytes transport therapeutic material directly into tumors or inflamed tissue. This seems especially promising for difficult-to-access conditions, whether lung disease or brain disorders such as glioma, stroke, and Alzheimer's. But the body's own leukocyte-based delivery systems are mostly "self-guided" and as such difficult to manipulate; attempts to guide macrophages combined with microrobots using magnetic fields, for example, are still in their infancy.[20]

If the aim is to transport cargo through the bloodstream, cells with very different specific functions are required. Erythrocytes (red blood cells) are the most abundant cells in the blood of vertebrates, serving to transport oxygen. Human erythrocytes have neither a nucleus nor organelles. The resulting capaciousness and flexibility makes them attractive for biomedical research as potential carriers, but erythrocytes have another advantage. Not every cell can move within the vascular system and thus into the body's deepest interior. The blood flow differs considerably in the arteries, veins, and capillaries. To move in this environment, cells must have enough propulsive force to swim against the stream and enough deformability to squeeze their way through the smallest capillary vessel. The use of erythrocytes as microrobots is under investigation partly because they can move through vessels measuring only half their own diameter, guided by ultrasound.[21]

Among the most extensively studied transport systems for biohybrid microrobots so far are bacteria. Bacteria convert chemical energy into movement and orient that movement with the aid of temperature, chemical attractants, oxygen, and pH values. They also react to light and magnetic fields, and being prokaryotic cells, have no nucleus. Thanks to these properties, they can be deployed as biohybrid "microbots," successfully tested *in vivo* for the first time in 2007. The researchers loaded a fluorescent gene onto nanoparticles to be carried on the surface of a *Listeria* bacterium. When the bacterium was injected into a live mouse, the gene was expressed.[22] Much research is currently exploring the manipulation of bacterial taxis, including a "termination switch" that it is hoped will dispose of the bacteria once their work is complete.[23]

Bacteria are usually propelled by flagella in a huge variety of forms. These outward-pointing, whiplike structures may be arranged at one or both ends of the bacterium body (monopolar or bipolar), single or multiple (monotrichous or lophotrichous), or distributed over the bacterium's entire surface (peritrichous). In the case of the helix-shaped spirochetes, the flexible bacterial body is itself wound about a bundle of flagella in an endoflagellum arrangement, so that the whole bacterium moves like a corkscrew. Depending on their number and distribution, the flagella can propel the bacterium clockwise or counterclockwise in twisting swim movements, described as a "run-and-tumble motion" that resembles "a sort of random walk, to explore their surroundings." In the run phase, the flagella rotate counterclockwise, leading to bundling and thus a constant motion, swimming "in a straight path." In the tumble phase, the flagella change their rotational direction and "unbundle," altering the bacterium's orientation. The direction of movement is controlled by receptors that probe environmental conditions favorable or unfavorable to the bacterium and if necessary can reorient it by changing the rotation.[24]

Flagella and cellular receptors also direct movement in some microalgae, such as *Spirulina* and *Chlamydomonas*, which interest researchers due to their "perception and driving skills."[25] The alga's flagella follow a certain choreography, rather like the breaststroke in human swimming: "a two-beat sequence of asymmetric deformations." It can vary the speed of its movement with "power strokes" and "recovery strokes." *Chlamydomonas* algae also "randomize their trajectories" by alternating between synchronous and asynchronous beating.[26] Finally, the swimming movement may be influenced from outside the organism, through phototaxis — the alga's movement in response to light. Set to work as a "microscale beast of burden," the alga is able to "transport microscale loads (3-μm-diameter beads) at velocities of ≈ 100–200 μm·sec-1 and over distances as large as 20 cm."[27]

Eukaryotic cells such as sperm cells, too, use flagella to move through space. Compared with algae, sperm cells have the advantage of being not only endogenous to the body and thus completely biocompatible, but also highly adapted to a specific biological milieu — they are "naturally optimized to swim in the female reproductive system."[28] This is a very demanding environment, marked by viscoelasticity, mucosal barriers, and the confined and convoluted anatomy of the fallopian tubes. Yet sperm are able to survive and retain their motility for several days in this difficult world. They swim at high speeds and generate highly modifiable undulatory movements with their flagella, applying a "slithering" mode close to a surface or a "swimming" mode along a wall.[29] Sperm motion includes "helical, hyperhelical, and hyperactivated movements; chiral ribbons; and slithering," the mode always determined by "the geometrical, physiological, chemical, and rheological stimuli in their environment."[30] In 2017, these properties were put to use in the "spermbot," a sperm-driven biohybrid—self-propelled, but guided magnetically—that could, at least *in vitro*, deliver a drug attacking cervical cancer cells.[31]

Motor and Temporal Regimes

We have encountered moving sheets of silicone and cells made of something that involves, but exceeds, organic matter. All of them have moved inside and in tune with the body or exploited the body's own forms of motion. The body is the natural model to which technology aspires; it is the constraint on intervention in its interior; and it is the obstacle whose organicness must be stretched and reworked by technology. As technology runs up against its limits, it persists in trying to shift the borders of the body by means of the body's own potentials.

Of all the borders that technologies try to exceed, all the things that human beings try to optimize in themselves and their environment, one may be the very emblem of our era: the organism's temporality. Bodies come into being and bodies die. Processes in the body obey their own temporal regime. They are perfectly synchronized, multidimensionally orchestrated, a delicate fabric of overlapping time scales. This may explain why the most recent research in the field has turned to a parameter very difficult to manipulate by nonsynthetic means—speed. The aim is to create "rockets" or "microjet engines," nanoscale synthetic motors able to move through the body at speeds far surpassing that of cells, motor proteins, or microorganisms.

For these rockets to be able to hurtle through the body's interior, they have to remain compatible with their natural milieu, human cells, and they have to run on nontoxic fuels—biocompatible and biodegradable substances.[32] Whether soft robots, microbots, or biohybrids, in the future, all of them will need to disappear into the body when their work is done. They must become part of its milieu, dissolving without residue into the fluid they exploited for their motion.

Matter is animated in these microrobots in a dual sense. On the one hand, they are animated in movement — in the imitation of nature in its essential form; on the other, they are animated in movement's cessation — in the fact that the end of the movement entails the end of the material, or rather must entail it if the robot and its intervention are to remain beneficial and not become a toxic presence. Paradoxically, in their motion, microrobots take the definition of life through motion to the point of absurdity, but in their stasis, they confirm that definition: their functionality depends on the cessation of their movement coinciding with the cessation of their material existence.

Motion marks the beginning and the end of organic time. We do not know if rockets in the body's interior will really work, but we do know that they operate on a time scale more closely akin to that of the twenty-first century observer than to the imperceptible, viscous flow inside our bodies. By controlling motion, technology sets about commanding time.

CHAPTER FIVE

Performing Organisms

Nothing much happened in the eighteenth century—from the point of view of the history of microscopy, that is. The seventeenth century had seen the invention of the microscope and the first access to a miraculous, previously unimaginable world. The nineteenth century would bring advances in microscope technology and the path-breaking discovery of cells and pathogens. Sandwiched between miracle and discovery, eighteenth-century microscopy had to make do with mere curiosity. In the history of science, the eighteenth century has thus seemed little more than an inevitable interlude, and for a long time, the microscopy of the period was largely disregarded.[1]

Yet "curiosity" is itself a remarkable thing: on the one hand, the desire to know or learn something; on the other, a strange or unusual object or fact. This chapter is about just such an object or fact, strange to its own era and alien to modern historiography, where it has passed unnoticed. For in the eighteenth century, animate motion was a curiosity. It was fascinating, perplexing, and—ever since 1675, when Leeuwenhoek announced what he had found in a drop of rainwater collected in a blue-glazed pot—carefully observed. Motion animated the miracle of an unknown universe suddenly appearing behind a glass-bead lens almost as tiny and delicate as what it revealed and the spectacle that the lens brought

4

4

forth: organisms made to perform for an audience. This particular strange and unusual object posed epistemic challenges that excited nothing less than insatiable curiosity.

Living Creatures in Rainwater

Not surprisingly, Antoni van Leeuwenhoek (1632–1723) was disconcerted when he looked through his microscope for the first time. Yet his famous letter to the Royal Society in London of October 9, 1676, begins with aplomb: "In the year 1675 I discover'd living creatures in Rain water."[2]

Leeuwenhoek presented all his investigations in letters — to the scholars of his day, including Robert Hooke (1635–1703), Nehemiah Grew (1641–1712), and Christiaan Huygens (1629–1695), and to the secretary of the young Royal Society, Henry Oldenburg (1619–1677), who made them available to an erudite public in the society's *Philosophical Transactions*.

Leeuwenhoek's celebrated 1676 letter, of which Oldenburg translated parts from Dutch into English for publication, is arranged like a journal of his observations of those "living creatures."[3] He scrupulously records the type of liquid and the circumstances in which he found and studied the organisms, discussing their size, appearance, and movement; he observes their propagation in standing water and calculates their amazing, constantly proliferating numbers. Leeuwenhoek's account presents research as an everyday activity and seeing as part of an attentive perception of the whole world around him. One can almost feel, taste, and smell it in his words.

Thus, on a rainy day, Leeuwenhoek gathers the water "running down from the house-top," ensuring that "no earthy parts" become mixed with it. Caught in a "clean glass" and rippling in "a stiff gale of wind," the rainwater shines in the cloud-dappled sunlight. The microscopist takes water from the well in his yard, even in summer "so cold, that you cannot possibly endure your hand in it for any reasonable time"; and during a visit to the seaside at Schevelingen, "the wind coming from Sea with a very warm Sun-shine," he instructs a swimmer to gather a sample from deep in the sea — telling him to first wash a new glass vessel carefully and then "tye the bottle with a clean bladder."[4] Another time, having long been interested in "the cause of the pungency of Pepper upon our tongue," Leeuwenhoek decides to soak some pepper in water. After three weeks of infusion, what he discovers is unsatisfactory for the taste buds, but all the more bewildering for the eyes: he sees, "to my great wonder, an incredible amount of little animals, of divers kinds."[5] The same occurs with infusions of ginger, nutmeg, or spittle and tooth plaque mixed with rainwater.[6]

The prodigious numbers and proliferation of these little animals in minute quantities of water face Leeuwenhoek with a smallness he finds difficult to capture in words. He tries to categorize the creatures by distinguishing them according to their size, appearance, and form of movement. The very first ones he describes moved by putting forth "two little horns," which "were continually moved, after the fashion of a horse's ears."[7] He observes a "motion of extension and contraction." The animals had "divers incredibly thin feet, which moved very nimbly," and "their body was also very flexible; for as soon as they hit against any the smallest fibre or string, their body was bent in, which bending presently also yerked out again." Other organisms "moved very briskly, both in a round and straight line"; a further type "exceeded all the former in celerity. I have often observ'd them to stand still as 'twere upon a point, and then turn themselves about with that swiftness, as we see a Top turn round, the circumference they made being no bigger than that of a small grain of Sand; and then extending themselves streight forward, and by and by lying in a bending posture."[8] Some "very small *animalcula* did swim gently among one another, moving like as Gnats do in the Air; so did these bigger ones move far more swiftly, tumbling round as 'twere, and then making a sudden downfall."[9] In seawater, Leeuwenhoek also finds a "blackish" little animal with "a peculiar motion, after the manner as when we see a very little flea leaping upon a white paper," while another seawater creature showed a "Serpent-like" motion. In the pepper infusion, he admires some with "a very pretty motion, often tumbling about and sideways" and others with different movements "no less pleasing or nimble."[10] Again and again, he is struck by the "incredible" swiftness of their motion.[11]

In the extensive sections of the letter not published by Oldenburg, Leeuwenhoek elaborates on the movements of the organisms, which he endeavors to identify and describe with the greatest precision, despite their minute size. In the pepper infusion, for example, he observes "with very great wonder" a kind of animalcule that is almost transparent: "'twas thus very troublesome to succeed in seeing 'em alive in the water. They moved with bendings, as an eel swims in the water; only with this difference, that whereas an eel always swims with his head in front, and never tail first, yet these animalcules swam as well backward as forwards, though their motion was very slow." He finds animalcules whose motion "was mostly all a-rolling, wherewithal they didn't much hurry themselves," or "all a-wallowing, on their back as well as on their belly."[12] After many days observing a growing profusion and diversity of animalcules, he finds a type that "had the figure of a pear" and another that looked more like dates, but no thicker than "a very fine sheep's hair." Their motion "was very curious, with a rolling about and a tumbling and a drawing of themselves together into a round."[13] In ginger water, finally, he finds creatures that move quite differently from the ones in the pepper water: "while the animalcules in the pepper-water went forward all winding-wise, these animalcules all advanced in jumps, hopping like a magpie."[14]

The movements are both wondrous and aesthetically pleasing:

The whole water seemed to be alive with these multifarious animalcules. This was for me, among all the marvels that I have discovered in nature, the most marvellous of all; and I must say, for my part, that no more pleasant sight has ever yet come before my eye than these many thousands of living creatures, seen all alive in a little drop of water, moving among one another, each several creature having its own proper motion.[15]

It Moves, Therefore It Is (an Animal)

What Leeuwenhoek discovered under the microscope had never
been seen before. Few had looked through the lens at all, and no
one at such a high magnification.[16] Reception of both the discovery and its author was correspondingly skeptical. Leeuwenhoek
did not belong to the academic community of European scholars
and did not speak their language, having mastery only of Dutch.
Conversely, he did not give them access to his own idiom — never
commenting on his characteristic practical talent in microscopy, his
tactile intuition, or his dexterity in grinding the tiniest lenses with
the greatest refraction.[17]

A few erudite advocates, including Reinier de Graaf (1641–1673)
and Constantijn Huygens (1596–1687), nonetheless recommended
him to the Royal Society, specifically as "a person unlearned both
in sciences and languages, but of his own nature exceedingly curious and industrious."[18] And thanks to his detailed accounts of the
production of samples and infusions, his chronology of observations, and his comments on the appearance, size, and number of
the animalcules, his contemporaries were able to participate in his
endeavors on paper and ultimately in practice. Leeuwenhoek used
their "virtual witnessing" to assure both himself and his addressees
of the credibility of what he had seen.[19] He had to work hard to be
listened to and to have the microscopic world seen.

Much could be queried or doubted regarding his discovery, but
there was one thing that Leeuwenhoek believed entirely beyond
dispute: that what he had observed were living creatures. It
remained a mystery *why* he found them in fluids that fell from the
skies or bubbled up from the earth, in fresh, half-evaporated, or
specially prepared infusions — what interested Leeuwenhoek was
the sheer smallness of the animalcula and their overwhelming numbers, which he estimated or calculated many times.[20] The question
of what exactly he had seen through the lens was not an epistemological one that seemed worth lengthy deliberation, but simply a
matter of magnitude. The transition from the world this side of the

lens to the world on the other side was seamless. Accordingly, his letter introduces the animalcula as tiny variants of the water creature that had first been described by Jan Swammerdam, "by him called Water-fleas, or Water-lice."[21] When Leeuwenhoek writes of "living creatures," "animalcula," and "living atoms"—often using the Dutch diminutives *dierken, diertgens, diertjes,* or *kleijne* and *seer kleijne dierkens*—he means very small animals, whether they can be seen with the naked eye or only through the microscope.[22] Some of them are not even nearly as large as "the eye of that little animal, which Dr. Swammerdam calls the Water-flea." Compared to a "Cheese-mite (which may be seen to move with the naked eye)," the creatures are as small as "a bee to a horse."[23]

The sole reason why Leeuwenhoek is so sure he is seeing animals is that they are moving. The forms of that movement, deciphered with such care and admiration, are no mere curiosity, but the very proof of their identity.[24] What moves is an animal—regardless of its size. At the time of Leeuwenhoek's discovery, toward the end of the seventeenth century, little could be said about life except that it moves. Movement, though, did make it possible to put life to a practical test by experimenting with life's end: death commences with the loss of movement, and what no longer moves is already dead. Thus, Leeuwenhoek added more pepper to his infusion and found that although the animalcula were still alive, "their motion was not so quick as when they were in plain well-water."[25] If pepper was replaced by vinegar, "the multifarious very little animalcules that were next the bottom, where the vinegar was, lay without any motion," and those farther away "became slower in their motion." After a while, "all the very little animalcules were dead."[26]

Among the authenticating eyewitnesses to whom Leeuwenhoek demonstrated his lenses and infusions was the pastor of the English congregation in Delft, Alexander Petrie. Petrie found the vinegar experiment particularly fascinating: "being desirous to see a proof whether those *animalcula* were indeed living animals, Mr Leewenhoeck by adding a very small quantity of vinegar to the same water, and putting it again into the same glasse-pipe, I did see those litle animals in the water, but they did not moove at all (being killed by the vinegar) which I beheld with admiration."[27] In a later letter, Leeuwenhoek tells Robert Hooke that he sees the animalcula "alive and exhibit as plainly to my eye as one sees, with the naked eye, little flies or gnats sporting in the air . . . for not only do I observe their progression, both when they hurry, and when they slow down, but I see them turn about, and stand still, and in the end even die."[28]

In 1677, Leeuwenhoek also saw animalcula in human semen, his own: small round bodies with thin, thread-like appendages. Over the course of many years, he continued to seek and find animalcula in the semen of any animal he could get hold of — rats, fish, oysters, and insects, to name just a few. Dissecting the ovaries and testicles of live animals, he asked what these animalcula were doing in the body. Was the animal inhabited by other animals? Where did they come from? Were they alive? Why did they move?

Leeuwenhoek's discovery of sperm contributed to a debate on the generation of animals that was seething at the end of the seventeenth century, about preformation, spontaneous generation, and epigenesis. Arguments in favor of the ovum's significance in generation had received new momentum shortly beforehand from the discovery of the Graafian follicle by Reinier de Graaf.[29] Leeuwenhoek conducted a veritable campaign against this dominant "ovist" position. He invoked not only his proof of the existence of sperm, but also, and particularly, their motion. Only because they are themselves animals, argued Leeuwenhoek, can the animalcula ensure the reproduction of animals.[30]

In a letter of 1685, Leeuwenhoek recounts his study of the semen of a dog in more detail. Finding that even after seven days, some of the animalcula in the semen "swam as briskly as if they had just come from the Dog," he infers "that these Animals, would have lived a much longer time, if they had been in the Uterus."[31] The semen remains alive in the female reproductive organs. Indeed, the animalcula "are proportioned by Nature" to move through the "narrow and wrinkled" passage of the female reproductive system, for their task is to carry "the semen into the Uterus."[32] This gives the lie to anatomical views according to which the semen never reaches the uterus, Leeuwenhoek argues.[33] The uterus, and not the vagina, is the destination of the animalcula, given that they swim only "forward in the Uterus"—unless they later swim back to the vagina, which would, as he remarks dryly, "be unprofitable ground."[34] To be precise, they moved forward five and one-third inches, the distance that Leeuwenhoek pointedly notes for the route shown in his illustration. He calculates that forty minutes will be required for the sperm to cover this distance.

The significance of the animalcules' movement is not only that it makes reproduction anatomically possible. Leeuwenhoek adds that one must assume animals have "a living Soul" even before birth, while still in the womb. If this is so, he reasons, "its a thousand times more likely, that the Soul which is in the *Animalia Seminis Masculi* should still remain there, changing only their outward shape." Because the seminal animalcule is an animal and consequently possesses a soul, Leeuwenhoek arrives at a conclusion with far-reaching consequences for the theory of generation: "I am also perswaded, that the living Soul which is in an Animal of the *Semen* of a *Cock*, does not loose it self in the parts of the Egg." The contrary must be the case: "the parts of the Egg pass into the living Animal; so that the Egg of the Hen serves only to nourish and feed the living Animal."[35] It is only due to the movement of the animalcula in the semen that, first, the animal can anatomically ensure its continued existence and, second, the soul can reinscribe itself into the next generation in the womb and thus reproduce the animal *as* an animal.

Still Lifes
The discoveries of the seventeenth and eighteenth centuries were lovingly displayed in cabinets of curiosities or *Wunderkammern*. Specimens and wonders were offered up to the gaze pillowed in little boxes, arrayed in cupboards, or suspended from the ceiling.

They were outlined in drawings and engravings, bound in laboratory books and magnificent, often colorful folios, and sent around
the world by a flourishing book market. Much work in art history and the history of science has been devoted to this universe
of books and images, to collection and exhibition as a component
of microscopic knowledge.[36] But whereas Leeuwenhoek's prolific
words breathed life into the world that he saw, in pictures it was
immobilized.

Contemporary illustrations represent the agility of the animalcula in little more than a few curved forms, serpentine waves,
or dotted lines. Swammerdam's image of the water flea, which
Leeuwenhoek cited as a comparison for his animalcules, indicates
its movement through the water at most through the slightly less
regimented crosshatching of the background (fig. 5.1).

Leeuwenhoek's "little animals between teeth" (fig. 5.2) are
dynamically elliptical in form (fig. A), move in gentle curves from
point C to point D (fig. B), or hint at a snaking motion (fig. F).
Compare this with the explanatory text in his letter of September
17, 1683. The largest type of animalcule, Leeuwenhoek writes there,
"had the shape of *A*. their motion was strong and nimble, and they
darted themselves thro the water or spittle, as a Jack or Pike does
thro the water." The animalcule in fig. B "spun about like a Top,
and took a course sometimes on one side, as is shown at *G*. [i.e., C]
and *D*." The third type, visualized in the unremarkable circles of
fig. E, can be compared with nothing better "than a swarm of Flies
or Gnats, flying and turning among one another in a small space."[37]

Again, compare image and description in the letter of December
25, 1702 (fig. 5.3). The bell-shaped animalcula shown in the right-hand
section "so stirr'd the round Cavity of their Bodies, that they put
the small parts of the Water into such a motion, that I could not
see those Instruments they used to produce the said motion." Leeuwenhoek continues: "I saw 20 of these Animalcula together, gently
moving their long Tayls and out-stretcht Bodies, they contracted
their Bodies and Tayls in an instant, and then softly extended them

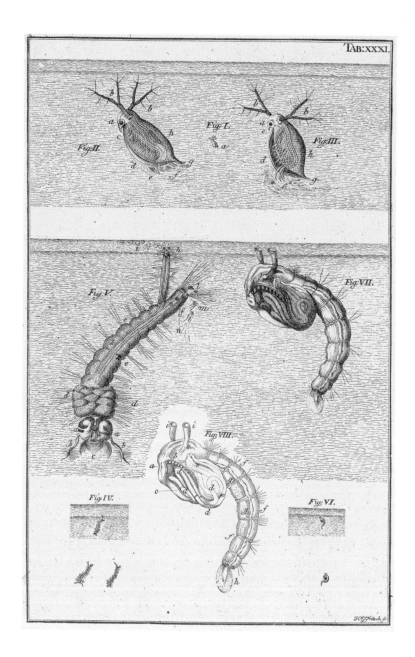

Figure 5.1. Swammerdam's illustration of the water flea (*Daphnia* spec.), in Swammerdam, *Bibel der Natur*, 1752.

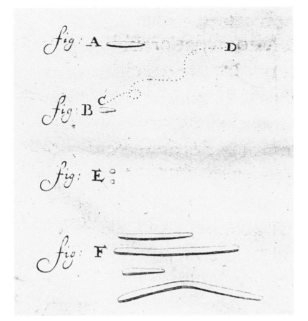

Figure 5.2. "Little animals" between the teeth. Leeuwenhoek, letter of 1683.

again; and this kind of motion they continued a great while, so that it was very diverting to observe them."[38] If the diverting spectacle of the wheel-like rotifers remains absent from the plate, that is not because Leeuwenhoek lacked the skill to reproduce it. Well aware that he could not draw, he engaged a draftsman, whose reaction he describes: "The Limner observing the Rotation of the Wheels, which always ran one and the same way, could not be satisfied with the sight, adding, O, that he could always see such a wonderful kind of motion." The wheels "were as thick beset with Teeth, like the Wheel of a Watch; and when these Animalcula had for some time exerted their circular motion they drew their Wheels into their Body, and their Body wholly into their sheaths, and then soon after thrust them out again with the aforesaid motion."[39]

The circulation of blood in young eels that Leeuwenhoek discovers in 1686 is likewise an event full of energy, a thrusting forward,

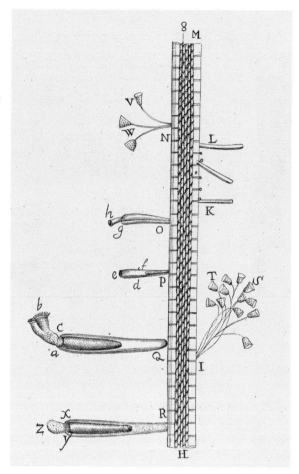

Figure 5.3. Animalcula around green weeds. Leeuwenhoek, letter of 1702.

pressing and pulsing, in movements "so quick one after the other, that with our Mouth we could not so quickly pronounce one Syllable after an other."[40] Leeuwenhoek observes something similar in shrimp, comparing the circulation to a "snowdrift propelled by a strong wind moving past our eyes," "as if we thought we saw through a large tear or the opening of a Window." He doubts that he "ever saw blood being propelled with such speed."[41] It was a sight that "did far exceed all the strange and pleasant ones, that ever mine Eye did behold before."[42] As Karin Leonhard has shown, this pleasure is petrified in the 1698 drawing of the blood's circulation (fig. 5.4). It has frozen into a carefully interlocking, static pattern of corpuscles and globules.[43]

Leeuwenhoek's visual world is "freed of all sensual pathos,"[44] the price of that liberty being that his imagery does not remotely reflect his sensual experience. Certainly, Delft lace appeared as thick as rope under the microscope, spiderwebs as solid as ramparts, salts as symmetrical crystals. But the ornamental quality of the pictures is something less than a depiction of those perceptions, and the urgency of the texts that describe the sensual dazzling of perception is something more than baroque flamboyance and obsessive detail. This moving world in miniature was one that took the microscopists captive: they could not confront it as mere observers. A different knowledge, another form of perception, existed in which the observer does not confront the world from the sidelines, powerless to portray it. This observer does not look at the moving world as if through "the opening of a Window," but stands right in its midst.

Lenses, Light, and Shadow

The lenses that magnified these diminutive things a hundredfold also played with light and shadow in the camera obscura and the magic lantern and projected insects into the room through the solar microscope. They made it possible to cast ceiling-high onto the wall what the senses could grasp only with difficulty and yet found so irresistible: living motion. The movements of beetles, fleas, and animalcula were put on stage, catapulted from light into dark through lenses and mirrors, transformed from the world of the microscopically small, invisible to the human eye, into the world of the spectator.

What could be seen only through the lens, in the ephemeral play of light and shade, has not found its way into our knowledge of

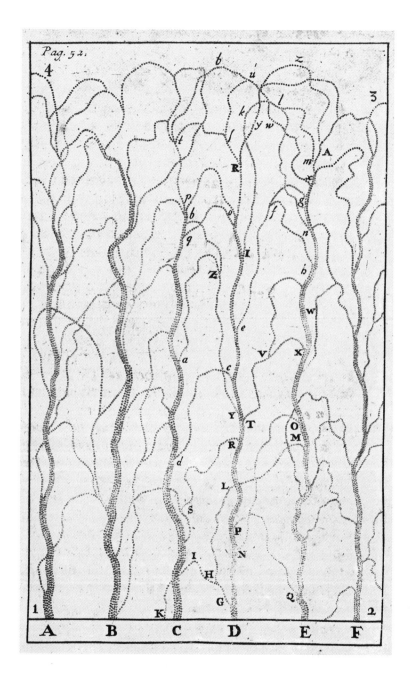

Figure 5.4. Blood vessels in the tail of a young eel. Leeuwenhoek, letter of 1698.

seventeenth-century and eighteenth-century microscopic explora-
tions. Whereas fine art appears as both a necessary and a worthy
partner of science, these projections have been regarded as a tran-
sient spectacle, an entertainment — designed to frighten or amuse
the uninitiated, untouched by scientists, and enjoyed by amateurs
only within the constraints of a gentlemanly practice. The motion
of the living world preoccupied the seventeenth and eighteenth
century sensually and epistemically, but this absorption was lost
to later generations, because projections are performative. Motion
resists capture in the form of images, and its significance is easily
overlooked.

In fact, starting in the second half of the seventeenth century,
new or constantly modified lenses and other optical devices spread
swiftly all across Europe. If we look beyond the technology of the
apparatuses, their origins, and the separation between content and
applications to approach projection instead as, literally, a process
of "projecting" from one place to another, of stepping from one
domain into another, then projection becomes a space of experimentation for new sensual experiences, defying all boundaries.[45] As
Marc Ratcliff has shown, collective viewing, in particular, became
crucial to the credibility of microscopy in the eighteenth century,
if only because standardized lenses did not yet exist.[46]

The fundamental principles of the camera obscura had been
known since antiquity. Sunlight enters a darkened room through a
hole in the wall or in a window shutter, and whatever is happening
outside that opening is projected onto the facing wall inside. The
projected image appears in motion, in color, and in reverse (upside
down and reversed from left to right). The closer the facing wall or
a stretched white cloth or paper is to the opening, the smaller and
more sharply focused the image becomes.

The moving projection was a phenomenon firmly rooted in the
experiential space of everyday life,[47] but its usefulness was harnessed for the first time by science, to study the stars. From the
thirteenth century on, astronomers used the camera obscura to

follow the course of solar eclipses. Even in the early seventeenth century, Johannes Kepler observed the heavenly bodies, driven by the *anima motrix*, from a portable tent he had designed himself in which he could trace the solar eclipse on paper and, by moving through the countryside, capture the whole of the sky.[48]

We know less about the uses of projection in more mundane settings.[49] As Laurent Mannoni points out, Leonardo da Vinci (1452–1519) seems to have been among the first to mention viewing not only the sun, but other outside objects with the camera obscura, and Gerolamo Cardano (1501–1576) wrote about portrayals of street scenes in his *De subtilitate* of 1550. The leap to spectacle appears to have occurred soon afterward, judging by an account given by Giovanni Battista della Porta (1540–1615) in the fourth edition of his *Magiae naturalis* in 1589. There, della Porta mentions what can only be called performances, either spontaneous or arranged, accompanied by music and with scenes correctly oriented with the help of mirrors. The Parisian monk Jean-François Nicéron (1613–1646) writes of Parisian camera obscuras open to the public, where visitors could watch people hurrying over the Pont Neuf and seagulls gliding over the Seine. Projection had found its own architectural place in the city's public life.

When Leeuwenhoek made his microscopical discoveries in the second half of the seventeenth century, then, projection techniques were already firmly part of European visual culture. As well as the camera obscura, this was due to the *laterna magica*. In the magic lantern, objects or scenes are painted singly or in sets onto glass slides and projected through the light (usually candlelight) cast by the lantern into a darkened room. Leeuwenhoek's reticence regarding his practices means we cannot say what exactly he looked at and how. What is certain is that he was a master of lenses, light, and shadow and that his milieu was not short of scholars of optics and skilled technicians who worked with projections in a huge variety of forms. Perhaps the most famous of Leeuwenhoek's contemporaries, the painter Jan Vermeer (1632–1675), introduced an unprecedented spatial perception into his pictures, a precision and perspectival geometry achieved by the use, or at least the knowledge, of the camera obscura. The two men were neighbors in Delft; they moved in the same artistic circles and guilds, and Leeuwenhoek was appointed executor of Vermeer's will, though no direct documentation of their acquaintance has survived.[50]

Christiaan Huygens, one of the foremost mathematicians and physicists of the seventeenth century, identified the rings of Saturn in 1658 with his telescope and one year later built a magic lantern, possibly the very first.[51] And Robert Hooke, who would later be charged with verifying Leeuwenhoek's discovery of the animalcula, published a letter in the *Philosophical Transactions* on August 17, 1668, outlining an apparatus that seems to have been a combination of magic lantern and camera obscura. Hooke describes projecting either a transparent object or a "living creature," which must be strongly illuminated "by casting the Sun-beams on it by Refraction, Reflexion, or both."[52] He adds the technical detail that in the case of "Living Animals," which cannot be turned upside down but which he clearly wishes to project with his device, two separate glasses will be required to invert the image.[53]

The visual power of projection, and especially its natural superiority to drawings in portraying motion, did not escape seventeenth-century scholars. In 1629, the French Jesuit Jean Leurechon (1591–1670) described the "pleasure in seeing the movement of birds, men, or other animals, & the trembling of plants shaken by the wind"; the camera obscura can "represent naïvely well what no painter has ever been able to portray in his picture: movement continued from place to place."[54] Nicéron, too, a celebrated painter of anamorphoses, realized in 1652 that the camera obscura could represent movement — "something that will always be missing from the pictures of the painters" — so that one seems "to see everything alive & flying on the white sheet."[55]

If projections of natural phenomena were still rare in the seventeenth century, or rather our knowledge about them is still scanty, their potential was known at the time and circulated within a European network that stretched from the Netherlands via the Royal Society to Franconia, from the streets of Paris to Bavarian monasteries, from scholars to the common people, from the seventeenth to well into the eighteenth century.[56]

Alongside the Netherlands, the German-speaking region of Franconia was another hub for the optics of motion. In its flourishing world of lens grinders, mathematicians, magicians, and monks, some of the most influential actors of late seventeenth-century and early eighteenth-century projection plied their trade: Johann Franz Griendel (1631–1687), Johann Christoph Sturm (1635–1703), and Johann Zahn (1641–1707).[57]

Johann Franz Griendel was renowned as a projectionist far beyond the borders of Franconia. From 1670 to 1677, he ran a public cabinet of curiosities in Nuremberg, along with a workshop in which he carried out optical experiments and manufactured and sold all sorts of instruments. Griendel's microscopical study *Micrographia nova* appeared in 1687.[58] Johann Christoph Sturm was first a pastor, then professor of mathematics at the University of Altdorf. In his two-volume *Collegium experimentale sive Curiosum*, published in Nuremberg in 1676 and 1685, he presented an illuminated drawing desk, arranged as a box within a box to increase the brilliancy of the screen. With the help of sunlight and mirrors, the image outside the opening in the wall is projected onto the table, where it can be conveniently traced.[59] This projection device also appears in the work of Martin Frobenius Ledermüller, which I will discuss shortly.

Another notable work is Johann Zahn's *Oculus artificialis telediop- tricus sive telescopium*, published in 1685, which became its era's standard work on optics. A professor of mathematics at Würzburg and the provost of the nearby monastery in Unterzell, Zahn published the works of his teacher Athanasius Kircher and corresponded with both Christiaan Huygens and Robert Boyle. *Oculus artificialis* covers everything relevant to optics, from theory to practice, from

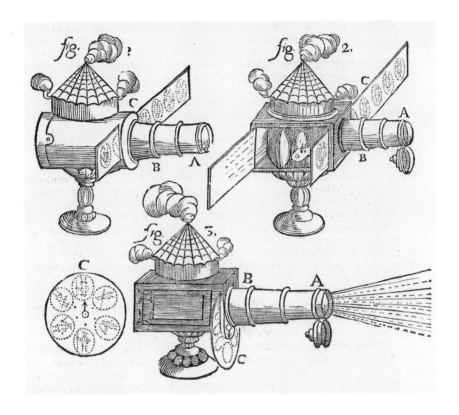

Figure 5.5. Drawings of a projection apparatus with various attachments that allow pictures painted on glass to be moved on slides or a rotating disk. Zahn, *Oculus artificialis*, 1702.

apparatuses to lens-grinding techniques. Even at this early stage, Zahn discusses projection with microscope lenses. He describes a device, called the "Lucerna megalographica," that is a "true form of microscope," since it greatly magnifies even the smallest objects. This "lamp" is also a magic lantern and at the same time a form of solar microscope, in that it can be worked with the rays of the sun deflected by mirrors as well as with a flame (fig. 5.5).[60]

Zahn carefully explains how he goes about projecting the smallest creatures. He names three technical variants for his magnifying lantern: using a shallow, hollow glass vessel that can hold liquids with the living animalcules; using two very thin, elongated glass plates; and using two leaves of some kind of waxed or oiled paper that are firmly glued together.[61] Common to all these constructions is that they function exactly like the glass pictures in the magic lantern and can be used in their place.

What this device brings to light can be described only in superlatives. Nothing at all, writes Zahn, is more curious to see than the images of the animalcula thus projected as they move around and run in every direction, presenting the "most delightful and athletic" performance to the inquiring spectator for maximal admiration and wonderment. When the animalcula, tiny flies, or almost transparent insects sense the radiance of the light or the heat of the lamp, it is with "the greatest astonishment" that one watches them, magnified, running around between the glass surfaces with their wings, hairy legs, claws, feelers, pincers, proboscises, and horns. If water containing various living "worms" is used, they appear moving on the wall like large and most wondrous snakes, to the enormous delectation of all.[62]

Enlightened Movement

With the Enlightenment, a new ideal of *Bildung*—in the sense of education and personal formation—arose, concerned with human beings' sensibility and capacity for perception. Optical instruments began to unfold a new, broad-based, and diverse potency, not only in scientific terms, but also socially. The instruments offered an appealing "optical complement to the tome-lined library," zestfully turning the prosaic knowledge of books into "theatrical productions."[63]

This process helped to fuel interest in the solar microscope. Unlike the classical microscope, but similarly to the camera obscura, the solar microscope produces its image not in the eye of an individual beholder, but on the wall inside a darkened room. Surrounded by gloom, the sun's light illuminates—in every sense of the term — the greatly magnified image of tiny objects or objects invisible to the naked eye.

During the Enlightenment, these microscopes perfectly fulfilled the ideal of a useful pastime and a "gentlemen's science," combining erudite entertainment with artful scholarship: they demonstrated microscopic knowledge in a public setting and presented it for discussion by an audience.[64] The devices form part of the social history both of microscopy and of the rococo period, when sociability, courtly life, and cultivated sensuality emblematized culture — and science as an essential component of that culture.[65]

Long belittled as scientifically marginal, at best a kind of toy, solar microscopy has attracted little scholarly attention. At most, the performances' spectacular dimension and the learned conversations that accompanied collective viewings have been mentioned.[66] Yet numerous solar microscopes have been preserved in European collections, testifying to the instrument's great dissemination and popularity in the eighteenth century.[67] Countless microscopy books, popular writings, and scientific treatises from the period, too, discuss the instrument and its use. Among the best-known microscopy books of the eighteenth century, reprinted multiple times and translated into many languages, were *Mikroskopische Gemüths- und Augenergötzungen* (Microscopic delights for mind and eye) by Martin Frobenius Ledermüller (1719–1769), *Insecten-Belustigungen* (Insect amusements) by August Johann Rösel von Rosenhof (1705–1759), and *Das Neueste aus dem Reich der Pflanzen* (The latest from the plant kingdom) by Wilhelm Friedrich von Gleichen-Rußworm (1717–1783).[68] These naturalists all worked in the city of Nuremberg — in the eighteenth century, a center not only of optics, but also of German scientific bookselling and book illustration in natural history and after London the most important place of publication in that field.[69]

Abundantly illustrated, these books followed the microscopical illustrations of the seventeenth century in presenting the frozen moment, often neatly framed by the circle of the eyepiece, as the central characteristic of microscopic imagery.[70] In the engravings

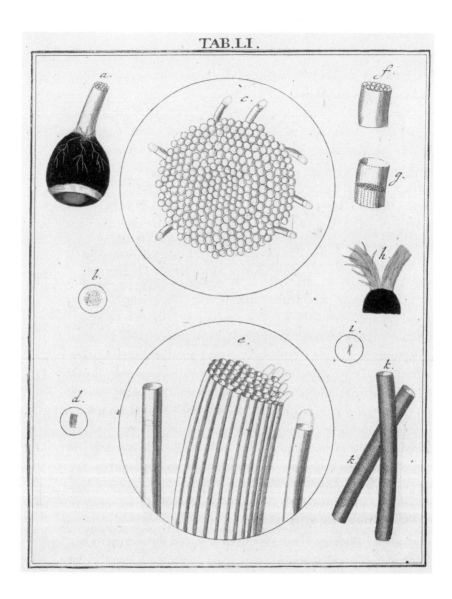

Figure 5.6. Optic nerve of a calf. Ledermüller, *Mikroskopische Gemüths- und Augen-Ergötzung*, 1761.

accompanying the great works on microscopy produced in Nuremberg, an iconographical proximity to ornamental patterns and
designs stands out. The images illustrating the works of Ledermüller, for example, are governed by geometry, symmetry, iteration, and planarity (fig. 5.6). In the high-contrast engravings of the
microscopy books, nature became image as a "visual utopia" where
growth found its echo in the formal language of ornamentation. This
is most clearly illustrated in the plates on polyps (fig. 5.7).[71]

But unlike these neat, orderly, and entirely pleasing depictions,
the microcosm actually glimpsed through the solar microscope's
lens faced its beholders with considerable visual and epistemic challenges. The solar microscope turned Leeuwenhoek's vivid verbal
evocations of tiny organisms and their enigmatic motion into real-
life images of gigantic creatures moving before the eyes of an audience — palpable and alive as hitherto only the painted presentation
of the projectionists had been. In the illuminated darkness, the
solar microscope offered a captivating, but confusing sensory experience of microscopic motion in actual living beings.

Comedy beneath the Cover Slip
What could be seen projected on the wall was a world of powerful
dynamism, abounding in chaos and unpredictability. The spectators
saw an unknown world full of ambiguity, of vague in-betweenness
and the neither/nor — one that called forth both enchantment and
"visual desperation."[72] The fact that this world was moving seems
to have contributed to the disconcerting effect in special measure.

The solar microscope in the darkened room thus, on the one
hand, conjured up incertitude and alterity, deception and illusion. It
projected, so to speak, the spectators' doubts about their own spaces
of perception and cognition onto the wall of the chamber, bathing
their misgivings in lurid light. It is for this reason that the world
of eighteenth-century microscopic projection was associated with
"visual quackery" or a "perceptual pathognomics"; it was puzzling,
bewildering, thrilling, "patched together like a showy monster."[73]

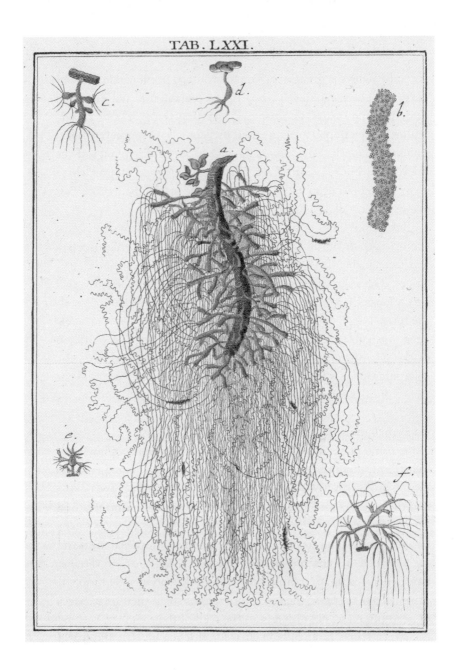

Figure 5.7. Polyp. Ledermüller, *Mikroskopische Gemüths- und Augen-Ergötzung*, 1761.

At the same time, the solar microscope offered microscopists a particular space of imagination within which they could begin to approach the cryptic and uncharted biological world. This applies especially to the understanding of living motion. Much was written in the period about the quality of movement, which was simultaneously perplexing and entertaining, clearly visible and resistant to visual representation. To grasp movement scientifically, epistemological exploration went hand in hand with aesthetics and sensory perception. New analogies and metaphors were needed — and these analogies and metaphors could be supplied by drama, theatricality, and the stage.

This becomes evident in the theatrical staging of a solar microscope session in Ledermüller's *Gemüths- und Augenergötzungen*, with its detailed information as to the precise drapes, seating, audience position, and dimming of the lights (figs. 5.8 and 5.9).[74] The creature that Ledermüller chose to illustrate the merits of his solar microscope was the flea.[75] Ever since it was portrayed by Robert Hooke in his *Micrographia* of 1665, the flea has been one of the very best-known objects of microscopic visualization, dramatically embodying the power of magnification. Logically enough, Ledermüller's account stresses size: the "white wall against which I am accustomed to cast the objects is more than four-and-a-half ells in height," yet even that "is not sufficient . . . to show the flea upright." Only by using different lens sizes can the flea be "contemplated in all its outward and inward parts, perfectly bright, clear, and transparent."[76]

The size of the insect, blown up to almost unimaginable dimensions, is one perception. But the insect projected for viewing is a living specimen, thus a moving image, and as the Romantic author E. T. A. Hoffmann was well aware, the perception of that movement is quite another matter. The protagonist of Hoffmann's 1822 tale "Master Flea" wonders at the uproar outside a darkened chamber:

> A glance into the room immediately showed young Pepusch the cause of the dreadful horror which had driven [the visitors] away. It was crawling with

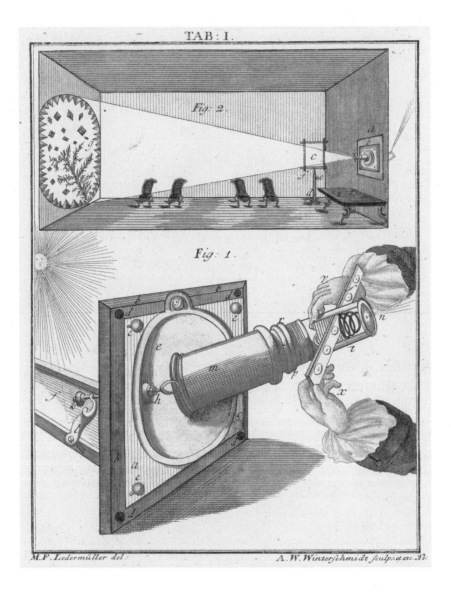

Figures 5.8 and 5.9. Guided through the solar microscope by a mirror, sunlight projects the object onto the opposite wall of the darkened room (5.8) or onto the transparent pane in a box (5.9). Ledermüller, *Nachlese seiner Mikroskopischen Gemüths- und Augenergötzung*, 1762.

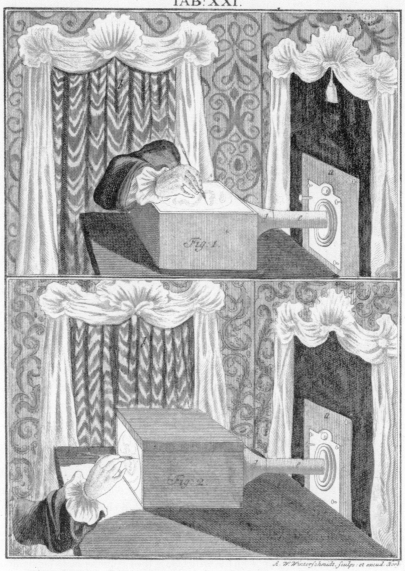

A. W. Winterschmidt, sculps: et excud. Norb.

life, filled by a pullulating mass of the most loathsome creatures. Lice, beetles, spiders, centipedes, all monstrously enlarged, stretched out their snouts
and strutted on long hairy legs; hideous ant-lions seized gnats and crushed
them with their jagged tongues, while the gnats tried to defend themselves
by flapping their long wings; among them coiled sea-anemones, paste-eels,
and hundred-armed polyps, and from all sides peeped infusoria with distorted human faces. Pepusch had never beheld anything more disgusting.[77]

The commotion in the room is the work of a solar microscope and
a flea tamer named "Leuwenhock," its description the work of an
avid literary imagination. Hoffmann uses metaphors of dramatic
art to capture the terrifying images generated by magnification
and movement. The tiny flea makes its entrance in a prodigiously
enlarged form, while an alarming number of different insects creep
and furl, wriggle and scramble through the pictorial space. Everything is jumbled confusion, everything is contorted and alive.[78]

"Fig. 1" in plate 75 of the *Gemüths- und Augenergötzungen* (fig. 5.10)
also shows an insect, a "muddy water insect" that Ledermüller calls
the Harlequin. On the lower half of the page, the creature is drawn
with fine and precise strokes. The other half (b, c, and d) shows its
anatomy and appearance in a high magnification and strong colors. In a series of tiny pictures along the top margin, we see the
life-size insect, like a red ribbon, wriggling and writhing, sidling
and coiling. Asterisks mark the direction of its movements. But the
sequence can offer no more than a pale reflection of what Ledermüller observed:

> I suppose I shall never confirm whether comedies are performed in the
> realm of the mud-dwelling insects, even though I have witnessed episodes
> and heroic deeds among them which could provide matter for the most
> beautiful plays on the stage; just as very recently I beheld with delight a
> single valiant polyp . . . on the battlefield as the victor over many hundreds of
> his enemies. But it is quite certain that among them lives a creature which
> resembles, in many a play, the droll figure of Harlequin. Its black head, its
> motley body, and its comical leaps and hopping twists and turns, some of

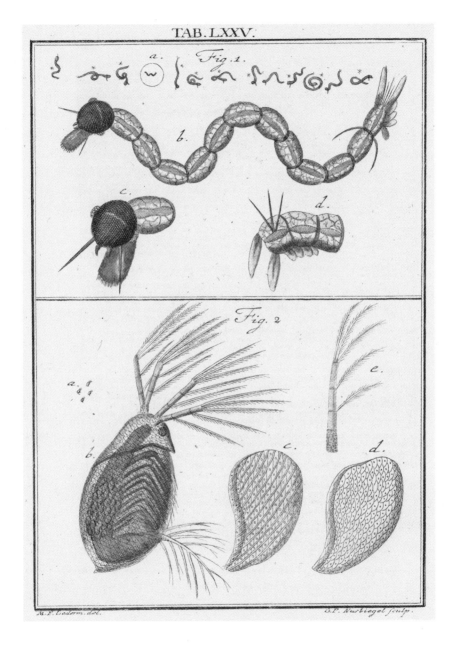

Figure 5.10. "Harlequin" insect. Ledermüller, *Mikroskopische Gemüths- und Augen-Ergötzung*, 1761.

which are signified with asterisks in the first figure of this seventy-fifth plate, have much similarity with that merry character of the Italian stage: for now the insect stands on its head ... now straight upright on its tail, adorned with broad feathery fins; now it lies stretched out, quite still, yet then recoils like a flash of lightning and takes a great forward jump like a snake. At times it rolls itself up like a bundle, peeps out maliciously with its black head, like a Scapino from his cloak, and then all at once leaps up high, finally bending like a drawn bow and proceeding quite circumspectly in this position, like an inchworm, on the water, upon which it is invariably, both at depth and on the water surface and at the bottom, capable of maintaining its balance, like a fish.[79]

Ledermüller's writings contain many such descriptions, and they are in no way inferior to Hoffmann's literary ones. Ledermüller writes of movements that are "uncommonly light and sprightly," proceeding at "an indescribable speed," "around in a circle" or "straight ahead,"[80] of insects that "row," "as swiftly as an arrow ... upward, downward, sideways, diagonally, and around,"[81] and of salts and crystals whose "swift operation and configuration" he has "oftentimes observed upon the wall," but that "can only be seen, and certainly not reproduced."[82] Under the lens, sea sand proves to be "a perfect show of aerial fireworks upon the water": "as if ignited," spheres can be seen spurting upward "exactly like a bullet or an iron cannonball," "flung around and jostled" as in a "comic play."[83]

Similarly, a view of the skin of a finger, drawn with all its "lines, fissures or crevices, pores and scales," reveals far more beneath the solar microscope: one can "see the exhalations of the hands, against the white illuminated wall, like the rising upward of some great smoke from the five fingers."[84] Whereas the structure may be meticulously registered by the engraver, the copperplate cannot seize the skin's vitality, its breathing. What the plates were unable to show, however, could be directly observed upon the wall. With solar microscopy, motion entered the auditorium on an equal footing with the audience. In the darkened chamber, the micro and the macro worlds confronted each other in a space created specially for the purpose — they were not, in the usual way, rigidly separated by the microscopic instrument.[85]

Science with the Sun

The motion made visible by the solar microscope inspired more than the imagination. It could also be used as a scientific argument, and Ledermüller did just that when he challenged Georges-Louis Leclerc Buffon (1707–1788), perhaps the era's greatest authority on natural history. Buffon's encyclopedic *Histoire naturelle générale et particulière* began to appear in 1749. In the second volume, Buffon set out his theory of generation, according to which nature consists of an infinite abundance of organic molecules that combine with the aid of an interior mold, organizing themselves into all the living forms of nature with enormous diversity and complexity. This notion put Buffon at odds with both the prevalent theories of generation at once: ovism in the tradition of Harvey and de Graaf, and Leeuwenhoek's animalculism. In Buffon's view, both sexes played a role in generation, and he identified seminal fluid (and thus generative material) in both sexes. Buffon's organic molecules, furthermore, are not specific to the sexual organs or even to reproduction — they also serve nutrition and growth.[86]

On the basis of his comprehensive new model of organic matter, Buffon joined forces with the skilled English microscopist John Turberville Needham (1713–1781) to investigate sperm through the microscope. He became the first to refute what Leeuwenhoek had observed so vividly: that the spermatozoa were animals. Buffon examined semen in human and other animal cadavers and in living dogs. Making consecutive observations over many hours, he saw that the sperm gradually became less mobile and concluded that the spermatozoa cannot be animals after all, because they do not move of their own accord. Buffon also argues that the appendage described by Leeuwenhoek is not really part of their body. After a few hours, it begins to shrink, so that the sperm take on the shape of small globules that then combine into new structures. The spermatozoa, therefore, are not animals, but "are properly the first assemblages of those organic particles so often mentioned; and, perhaps, they are the constituent particles of all animated bodies."[87]

Ledermüller strongly disputed Buffon's view. To counter the assertion that sperm were not living creatures, he invoked the solar microscope, through which he had observed their movements with great exactitude.[88] Using a Cuff solar microscope, Ledermüller explained, he had performed "hundreds, indeed thousands, of experiments."[89] Upon his very first "glance into the glass," he could see the spermatic animals in motion and "equipped with little tails"; they "swam about like young frogs, sprightly and fresh."[90] If Buffon had seen no movements, the sperm he had used must have already been "too old and glutinous"—for overdry sperm was like a "sticky mass" that immobilized "the little tails as if glued down."[91] Alternatively, Buffon might quite simply not have used a solar microscope at all.[92]

Almost exactly a century after Leeuwenhoek's letters, then, the question of whether spermatozoa were animals had lost nothing of its urgency. Their animality began to come seriously under fire from the mid-eighteenth century, when it was contested by such distinguished naturalists as Georges Buffon and Carl Linnaeus.

Ledermüller's position was supported by the nobleman Wilhelm Friedrich von Gleichen-Rußworm, who turned to microscopy in the wake of Ledermüller's *Gemüths- und Augenergötzungen*.[93] After his celebrated *Das Neueste aus dem Reich der Pflanzen* of 1764, von Gleichen-Rußworm published *Abhandlung über die Saamen- und Infusionsthierchen* (Treatise on the spermatic and infusorial animalcules) in 1778. In this important volume—rich in observation, detail, and innovative techniques—von Gleichen-Rußworm, like Ledermüller, frequently compares the spermatic animals to frogs. Just as hatching tadpoles "present the true spectacle on the large scale," on the small scale the sperm, under magnification, permit not "a moment's doubt" that "these little animals have a movement that is independent (originating from themselves) or voluntary."[94]

Lazzaro Spallanzani (1729–1799), an Italian priest and professor of natural history at the University of Padua, agreed with Ledermüller that only empiricism was capable of resolving the issue of sperm movement. Again like Ledermüller, Spallanzani argued that one reason why Buffon had failed to identify the movements of the spermatic animalcules was his use of a compound microscope instead of a single lens, since the compound microscope could not achieve the required resolution.[95] Spallanzani went much further than his predecessor, however, staking a claim to systematic investigation and methodologically informed observation. In *Opusculi di Fisica animale, e vegetabile* of 1776 (translated in 1799 as *Tracts on the Nature of Animals and Vegetables*), he devised numerous experiments to corroborate the animality of spermatozoa physiologically. If Linnaeus, for example, believed that the "seminal vermiculi" are "only inert molecules, swimming like oil in the seminal fluid, moving and darting in various directions, as they are agitated or heated by the temperature of the fluid," this was only because he had looked at "Lewenhoek's animalcula" far too hastily.[96] To rule out the idea that the spermatozoa glide passively on the surface of a liquid, Spallanzani focused his lens at different depths of the seminal fluid. He found spermatic animalcules in all of them, and they were all moving: "wherever the focus of the magnifier penetrated, I saw the motion of vermiculi."[97] Spermatic animalcules, then, must have "a spontaneous motion," for which they make use of their tails.[98] Spallanzani's "vermicules" move in two different ways, "one oscillatory, from right to left, and from left to right, curving the appendage from one side to the other," and one that is progressive, as "the vermiculus transports itself" by oscillation. As a result, they "not only swam horizontally, but rose and sunk in the semen, as fishes do in water."[99]

This discussion of movement is crucial to Spallanzani's onslaught on Buffon, because it leads him to conclude that Buffon did not actually see spermatic vermiculi at all, but confused them with a different kind of infusorial animalcula, which formed only after some time in the seminal fluid.[100] When Spallanzani expands on Leeuwenhoek's test of life *ex negativo*, tracking the creatures' death under the influence of tobacco, sulfur, or electrical sparks,[101] a systematic method and a strict observational regime lead him to the same conclusion as Leeuwenhoek's — backed up, indeed, by "conclusive evidence": "I esteem the facts I have related sufficient authority to bestow upon them the name of animals. The spontaneous motion, and the contortions of the body, by means of which they move from one place to another, are characteristics sufficiently decisive of their animality."[102]

Moving in Company

Animality was thus the core characteristic of spermatozoa — as it was of all the other "little fish," "protozoa," *Urtierchen* ("primordial animalcules"), "water insects," and "little serpents" that had fascinated the microscopists. Increasingly termed "infusoria" from midcentury, microscopic organisms continued to be studied throughout the eighteenth century in what was very far from being an established domain. Indeed, so confusing was the field that Linnaeus classified the infusoria in the genus of "*Chaos*," in the very last phylum of the animal kingdom under the class of *Vermes*.[103]

If this animality of the tiny organisms was defined by their movement, what exactly was the nature of that movement, which Spallanzani called "spontaneous" and Ledermüller likened to an actor moving on the stage? What made some motions a sign of vitality and others not? In his 1778 treatise on the spermatic and infusorial animalcules, von Gleichen-Rußworm distinguished "vital movement" from "mechanical" or "mechanical-organic" movement. The latter, he writes, can be found everywhere, being "a general law of nature."[104] Living or vital motion, by contrast, can easily be discerned by an "eye accustomed to looking through the microscope." One can neither see "something animated" in chemical liquids nor mistake the moving particles in plant juices for animalcules, because particles "are pushed forward straight ahead," whereas infusoria "travel in wavering twists and turns" and "move out of each other's way."[105] Animated movements are an "unmistakable sign of animality," because they are "unexpected" and "free actions."[106] Unpredictability, freedom, and voluntariness here become the hallmarks of living motion. Whereas Ledermüller described this feature in terms of a performer's agency, von Gleichen-Rußworm frames it within social behavior more generally. The animalcules, he writes,

> have the power to stop in the midst of their journey, to throw or turn themselves over and from side to side . . . to take a new path; the capacity to resist the mucous particles obstructing them, moving these and pushing them away; the deliberation to contract their body . . . when they encounter a narrow passage between two mucous particles . . . to twist . . . finally, to reveal passions when they assemble in troops, move off together in societies, and part again . . . or come together in pairs.[107]

Occurring in such large quantities in the smallest possible space, the infusoria displayed a social behavior that made them resemble the familiar animals of the macroscopic world. In *Descriptions et usages de plusieurs nouveaux microscopes* of 1718, Louis Joblot (1645–1723), professor of mathematics at the École Nationale Supérieure des Beaux-Arts in Paris and a renowned microscopist, discussed many new infusoria, which he almost entirely consistently calls "animals" (occasionally also "fish" or "insects").[108] As well as describing their multifarious movements, Joblot frequently explains how, despite their enormous agility and vivacity, they act together. He observes that two animalcula, both having moved along a straight line, now "leave between them a space too small for a third, marked L, to pass through it; the latter, caught and compressed between the two, stretches out and bends to escape toward M" (fig. 5.11).[109]

Although the animals are "swimming at great speeds . . . often brushing against each other," he sees them "sometimes stopping for a moment one facing the other," "changing their shape according to the manner in which they encounter each other," or he notices one of the "little oysters" moving "past a group of several others

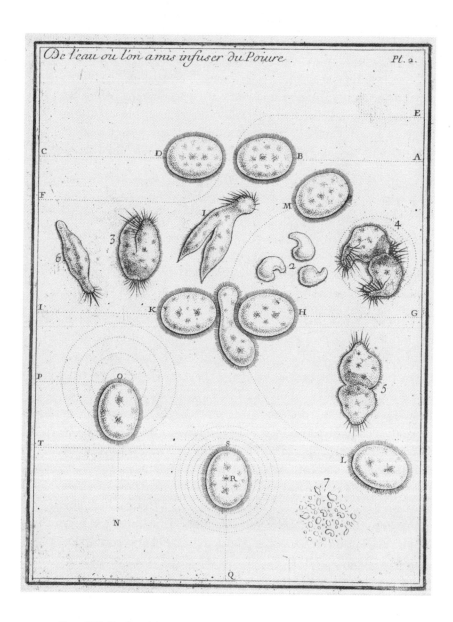

Figure 5.11. The three infusoria positioned in the exact middle of the image show the behavior described by Joblot. The infusorium L, now squeezed between infusoria K and H, is moving along the dotted line from L at the lower right up to M. Joblot, *Descriptions et usages*, 1718.

without shaking them; which indicates that they have not touched"
or that they "harmonize the movements of their bodies with such
precision that the convexities & concavities of some find an exact
response in those of the others."[110] Joblot compares two of the
creatures clinging together to "two turtle doves or two pigeons,
male & female, caressing each other."[111] Finally, he describes them
"appearing together as do a number of dancers who take pleasure in
entertaining a company." Even while swimming, they know how to
"stay out of each other's way, as do dancers appearing together in a
ballet scene."[112] The artistry of this spectacle surpasses everything
from the most popular to the most elegant artistic enjoyments that
early eighteenth-century Paris has to offer: "I do not believe that the
entertainments at the Comédie, of the Opera in all its magnificence,
those of the tightrope walkers, the tumblers, & the bear fights that
we can watch in this superb city, may be preferred to it."[113]

Pushing the Boundaries of Life

The seventeenth-century discovery of the animalcula revealed a
world that seemed just as animated as the world this side of the
microscope's lens. Whether in raindrops, mud, or bodies, whether
in the animals' transparent interior or in their external wheels and
paddles, movement was everywhere. The microscope had made it
visible for the first time, and the solar microscope brought it into
the world of the eighteenth-century spectator. What the microscopists saw left them in no doubt that the movements of life were
more than physical and chemical ones. If Ledermüller and Joblot
compared the motion of insects and infusoria to the Italian *comme-
dia dell'arte* and the French opera, this was not only because their
movements were just as comical, preposterous, or heroic as those
to be found on the stage, but because they showed such surprising,
unsuspected diversity and social complexity.

At the same time, Marc Ratcliff has shown, the framework of
discussions about microorganisms had changed considerably from
the seventeenth to the eighteenth century. Microscopy, first, moved

out of the private sphere and the discovery of invisible life into the wider world. It became a shared social experience of something that was only just visible, but whose existence (if no more than that), once viewed collectively through solar microscopes, became indisputable. Second, microscopy left the narrow confines of erudite correspondence and the academies: it was introduced, explained, and disseminated in the numerous treatises, compendia, and instruction manuals that appeared throughout the century. It thrived because those pursuing it were unrestricted—though also unsupported—by institutional frameworks. Finally, it left the space of optics, in which it had long held such a privileged position, and increasingly became part of a much more comprehensive experimental system. Microscopists such as Spallanzani, Ledermüller, von Gleichen-Rußworm, and Joblot devised every imaginable experiment to test their hypotheses, tinkering with gossamer-thin pipettes, mobile swivel arms, and holders for the specimens or vessels to store them in liquid. And once Abraham Trembley (1710–1784) succeeded in sending living specimens by post in the 1740s, careful experimentation meant not merely standardizing lenses, but also repeating observations using identical specimens.[114]

The passage of microscopic animals into the visible, shared, and experimental world set the stage for the nineteenth-century investigations to come. Above all, every new discovery about microscopic life and its astounding capacities posed new and fundamental questions about life in general. In 1744, Trembley announced he had seen the freshwater polyp regenerate its own body when cut into pieces; in 1765, Horace Bénédicte de Saussure (1740–1799) observed the division of infusoria; and increasing numbers of experiments on spontaneous generation fueled the debate as to whether life could arise out of nonliving matter or only out of similar organisms and their eggs.[115] Given these phenomena, seen so clearly through the lens, where exactly was the frontier between human and animal, animal and plant, dead and alive? Or more to the point, where was that frontier to be drawn in the future?

CHAPTER SIX

From Animal to Activity

When he discovered the animalcula in the late seventeenth century, Leeuwenhoek had no hesitation in deriving their animality from their ability to move. Yet from the eighteenth century on, the well-established triad of moving, living, and animal began to attract doubts — for the motion that became manifest in the world of infusorians defied all conventional demarcations. It was restricted neither to a particular creature nor to a particular habitat, nor did it belong to a single taxon, as Ledermüller's notion of the "infusorial animalcule" had suggested and the genus *Chaos* disproved. Such motion posed fundamental questions about the nature of the living. More than that: it became the central criterion by which the concept of life was renegotiated in the nineteenth century.

As the century began, everything that lived was an animal. By the end of the century, life resided in formless matter. Movement was still the mark of life, but now it epitomized life in both animals and plants, in cells and living substance, amoebae and protoplasm; whether outside or inside bodies; whether as a feature of beings or a behavior of matter. The investigation of movement had given rise to a new epistemic challenge: the key to understanding motion in an animal was the form of its organization, but in formless matter, motion was found in the becoming of form. Hand in hand with this shift came a profound visual challenge: whereas animal motion could be observed simply by watching animals move (even if through a lens), motion in formless matter had to be brought about and made visible in a single act.

Animality in Question

Leeuwenhoek consistently located his animalcula in the animal kingdom. Accordingly, he assumed that their locomotion was like that of animals. Where he failed to find feet, filaments, horns, or legs as locomotive organs, he simply concluded that these were too small to be seen even through the single lens.[1]

When August Johann Rösel von Rosenhof described the "globe-animal" *Volvox* in his periodical *Monatliche Insektenbelustigungen* (Monthly insect amusements) in 1755, he found an entirely new and curious form of locomotion: "It moves all over the place; it can roll away like a ball, spin like a top; or slide along lengthways without rotating in the least."[2] Rösel also suspected that this green alga was moving by a previously unknown mechanism that required none of the usual locomotive organs: "Namely, I consider the delicate and transparent little warts with which it is studded to be so many little hollow tubes, which the animal can open or close as it pleases and through which it can draw in and expel water or air," thus pushing itself forward, backward, and about its own axis.[3] The little globes showed no internal differentiation, "neither heart nor stomach." But because they moved independently, rather than being moved, "I naturally enough struck upon the thought that these little round bodies must belong to the living creatures."[4]

Another of Rösel's discoveries was the "little Proteus." This microorganism, which Rösel named after the mutable sea god, changes its form entirely through motion. Its internal substance of "transparent grains" constantly transforms its outer shape (fig. 6.1), so that the Proteus "now looked like P, then its antlers were lost and it changed into Q, from Q into R, from R into S, and from S into T; but at last it turned back into a ball that had a neck, and from this neck countless fine grains flowed out, so that the ball

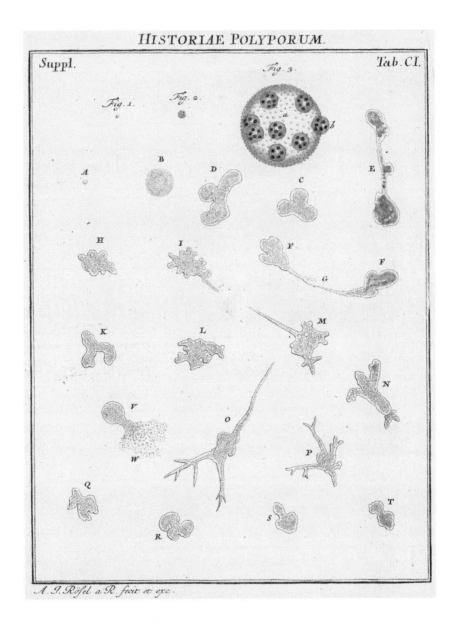

Figure 6.1. The first known portrayal of an amoeba. Rösel, *Der monathlich-herausgegebenen Insecten-Belustigung*, 1755.

resembled a fire-spitting grenade."[5] Rösel's Proteus was an amoeba, an amorphous organism formed of a granular substance enclosed by a membrane. On the one hand, it had no visible internal organs or differentiation like that of animals; on the other, despite lacking any locomotive organs in the real sense, it was capable of motion, a quality reserved to animals alone.

Organisms such as the amoeba also prompted the French naturalist Félix Dujardin (1801–1860), versed in geology, botany, and zoology alike, to study microorganisms. As he noted in his 1835 paper "Recherches sur les organismes inférieurs," it would "appear indeed quite strange" to declare these organisms animals, given that they have "no form of their own" — in other words, to speak of an animal even when an organism is devoid of any shape or internal organization and has neither digestive or reproductive organs nor anything even remotely resembling a vascular, nervous, or respiratory system. Although the infusoria clearly lack all visible interior, writes Dujardin, researchers since Leeuwenhoek have simply inferred such an organization "by pure analogy" with higher animals.[6]

That tendency is especially striking in the work of Dujardin's contemporary, the Berlin zoologist Christian Gottfried Ehrenberg (1795–1876), renowned taxonomist of thousands of specimens that he collected on expeditions reaching from Siberia to the Nile. In *Die Infusionsthierchen als vollkommene Organismen* (The infusoria as perfect organisms) of 1838, Ehrenberg recorded his (in fact erroneous) observations of the infusorians' internal organization, their digestive and sensitive organs, their means of locomotion — in short, what he considered to be "their perfect animal formation."[7]

Despite their diametrically opposed positions on the infusoria — as animals and as organisms fundamentally distinct from animals — Ehrenberg and Dujardin were of one mind when it came to accusing each other of being excessively fanciful. Dujardin described the attempt to identify in formless matter an organization analogous to that of animals as a self-indulgent relish in giving "the marvelous a role to play" in the new microscopic world;

Ehrenberg argued that ever since its discovery, the microscopic world had "often been portrayed by the pen of easily moved and fantastical authors as a monstrous realm of spirits, full of . . . unparalleled forms, some ghastly, some bizarrely distorted, not quite alive yet not quite lifeless."[8]

It is worth remembering that Hoffmann's story "Master Flea," featuring the flea tamer "Leuwenhock," had appeared not long before, in 1822. But the literary description of microorganisms caused less unease than the dawning insight that the creatures refused to be corralled along the established "highways of thought," as Frederick Churchill has put it.[9]

The Taxonomy of a Single Cell

By attributing an internal organization to the infusoria, even against all appearances, it was possible to regard them as complex organisms that displayed numerous other animal functions apart from motion. This "natural analogy" between infusorians and higher animals remained persuasive well into the nineteenth century, not least because the questions of sexual reproduction and the function of the cell nucleus in infusoria were far from having been resolved.[10]

In parallel, the cell theory that began to emerge at midcentury gave rise to the converse claim: not that the infusoria were complex organisms, but that a complex organism could consist of a single cell. This argument was articulated in the identification of the infusorians as unicellular animals and various attempts to tame them taxonomically.[11]

As early as 1817, the paleontologist and zoologist Georg August Goldfuss (1782–1848) coined the term *protozoa*, literally "first animals": "The animals of the first and lowest class are the protozoa, the spermatic animalcules of the primitive animal [*Urthier*]. Determinate directionlessness [*determinierte Richtungslosigkeit*]—that is, motion apparently at will, mostly by turning about their own axis—and a spiraling progress like planetary satellites, is the first act of animal life to reveal itself."[12] Drawing on Goldfuss's terminology, Carl Theodor von Siebold (1804–1885), in *Lehrbuch der vergleichenden Anatomie der wirbellosen Thiere* (Textbook on the comparative anatomy of invertebrates) of 1848, became the first to define protozoa as single-celled organisms, specifically as "animals in which the various systems of the organs are not sharply separated out, and whose irregular form and simple organization can be reduced to one cell."[13]

This position found few adherents in the mid-nineteenth century, a notable exception being the anatomist and physiologist Albert von Kölliker (1817–1905).[14] A broader shift came about only with the work of Otto Bütschli (1848–1920) in the 1870s. This confirmed the identity of the cell nucleus in infusorians and higher animals, a key presupposition for the cellular interpretation of infusoria. In 1876, Bütschli's comprehensive *Studien über die ersten Entwicklungsvorgänge der Eizelle, die Zelltheilung und die Conjugation der Infusorien* (Studies on the first developmental processes of the egg cell, cell division, and conjugation of the infusoria) addressed the reproduction of Ciliophora.[15] Bütschli realized that conjugation, in which two infusoria join together, is not analogous to sexual reproduction as found in higher animals, but is a form of reproduction specific to unicellular organisms. In it, the small cell nuclei of the ciliates he studied (micronuclei; he calls them "nucleoli") divide and give rise to new macronuclei in the next life cycle, a process during which they exchange cell material. Bütschli "acknowledged the nucleoli as true cell nuclei" and was able to give "sound evidence" that the nucleus in infusoria forms "through the growth of a nucleolus."[16] In addition, his detailed observations of the events during cell division (mitosis) in chick embryos, presented earlier in the treatise, underpinned the functional and morphological identity between the cell nuclei in infusoria and those in higher animals. This pioneering research on conjugation and the cell nucleus brought an end to the ongoing debate about the allegedly complex structure of infusoria, their sexual organs, and their reproduction.

Following on from that, Bütschli's 1887 contribution to the monumental animal taxonomy *H. G. Bronns Klassen und Ordnungen des Tierreichs* made unicellularity the cornerstone for the taxonomy of microorganisms.[17] He established the kingdom of protozoa — including the infusorians, sarcodina (equating approximately to the amoebae), algae, and sporozoans and excluding bacteria and unicellular plants — all on the basis of a nucleated cell. Bütschli's taxonomy thus turned Linnaeus's genus *Chaos* into one of nature's kingdoms, marking "the end of an era and the beginning of a distinctive field." His "unicellular model" provided the point of departure for the new discipline of protozoology to flourish, especially in Germany.[18]

Bütschli's pivotal work did not, however, settle another controversy sparked by the infusoria. The debate over whether infusorians, like higher animals, showed an internal organization had cast doubt on something hitherto taken for granted, the infusorians' animality. Whether creatures constructed so simply could fulfill the same complex functions as animals did was one question, but no less pressing was the counterquestion it entailed: If the infusoria really were so simple in their organization, what distinguished them from plants? In other words, were infusoria animals, plants, both, or neither?

The division of the living world into animals and plants had seemed the most obvious of all classifications for centuries, but to the botanist and algae researcher Friedrich Kützing (1807–1893), it was nothing more than a "dogma." "Every observer of the lower organisms," he asserted in 1844, "knows it is not possible to draw clear boundaries between the lowly beginnings of the plant and animal worlds," and he adduced a whole series of contemporary studies to prove as much.[19]

In the 1860s, the fundamental split into plants and animals was, as Lynn Rothschild has shown, attacked by several taxonomic classifications at almost the same time. These resolved the plant/animal dilemma by assigning the unicellular infusorians to a third kingdom: Richard Owen's Protozoa, John Hogg's Protoctista or Regnum

Primigenum, Thomas Wilson and John Cassin's Primalia, and Ernst Haeckel's Protista. In each case, the infusorians constituted a kingdom of their own alongside the kingdoms of plants and animals.[20]

The difficult and contested taxonomies of infusoria show scientists grappling for new categories in view of a world that is eluding the established order. For the present purposes, though, taxonomy is interesting less for its constant reshuffling of species between classes and kingdoms, which continues to this day. If the infusoria could not be pinned down as animals or as plants, perhaps not even in a kingdom of their own, and as the new taxonomies shifted the definition of life from the animal, via the plant, to the unicellular organism, everything now revolved about the cell. What, then, *is* a cell?

Elementary Organisms

Historians and biologists have long since shown that there is neither *a* cell, nor *a* cell theory, nor *a* history of that theory. They tell the multifaceted and tangled story of a concept that biologists from a very early stage described as a "misnomer ... merely an historical survival of a word," and as "a highly abstract concept of minimal content." For historians, the cell has been less a clear-cut concept than a metaphor and a "search image" to help them find particular features.[21] In the following, I do not dwell on the much-studied notion of the cell as an anatomical unit and a globular structure or on its being made visible through microscope technology.[22] My focus here is the designation of the cell as, to quote the botanist Matthias Schleiden (1804–1881), a "peculiar little organism."

Studying plants as entities composed of simpler units was a common approach among botanists of the time. In Schleiden's view, this simple unit was the cell — and that cell was an organism within the organism, which led a "double life," existing both "quite independently" for itself and indirectly as an "integrating component of the plant."[23] The zoologist Theodor Schwann (1810–1882), in his "cell theory," postulated only "one general principle for the formation of all organic productions," that principle being "the formation of cells."[24] For Schwann, then, the cell was above all a principle of development, common to plants and animals alike.

The notion of the cell as an elementary organism, a living unit, and a developmental principle marked a transition from designating the whole of a plant or animal as an organism to designating a tiny component as an organism itself. This meant that every cell, "every elementary part, possesses a power of its own, an independent life," and develops independently of the organism as a whole. In fact, Schwann locates not only development, but all physiological processes on the level of the individual cell, rather than that of the organism in its integrity: "the cause of nutrition and growth resides not in the organism as a whole, but in the separate elementary parts—the cells."[25]

Thus, cells became elementary living particles, the "ultimate agents of life," the fundamental, smallest living unit common to the animal and vegetable kingdom.[26] Whether in single-celled infusoria or highly developed organisms, the central physiological processes—respiration, nutrition, reproduction—were displaced into the individual cell. But the concept of the cell as elementary organism did more than shift the borders between living parts and wholes. Once life was attributed to tiny parts, attention turned from the structures in which the organism is organized toward the substance of which it consists.

Moving Fluids

Organic matter had been the subject of a rich imaginative tapestry since the end of the eighteenth century. Numerous terms circulated in Europe, from *Gallerte* (jelly), *Zellstoff* (cellulose), and *Grundschleim* (basic mucus), to *cells*, *globules*, and *Körnchen* (granules), to *tissue cellulaire* and *tissue muqueux*.[27] They all designated some kind of amorphous, mucous, granular, more or less fluid substance.

John Pickstone identifies a "coagulation paradigm," dominant from the mid-eighteenth to the mid-nineteenth century, in which animals and plants were composed of globules and the coagulation of fluids gave rise to tissue, especially that of blood and muscles.

German physiologists, in particular, considered this amorphous living substance to be "the universal material basis of life."[28]

The multifarious motion that took place in this organic matter was often regarded as its central property. In 1774 and 1811, respectively, Bonaventura Corti and Ludolf Christian Treviranus observed flowing movement in plant tissue.[29] Motion was also deeply significant for epigenetic notions of development, in which new forms arise not out of preexisting structures, but successively out of initially homogeneous matter. Thus, in Caspar Friedrich Wolff's mid-eighteenth-century epigenetic theory, organic structures are formed in an alternating play of flow and standstill in the nutritive fluid. The influential German physiologist Ignaz Döllinger identified motion as the origin of every formative momentum in the organism. It was in the alternation between a rigid order and a temporally variable one, in the metamorphosis of animal substance between movement and rest, that he found generation, nutrition, secretion — indeed, all physiological processes.[30] For Carl Friedrich Kielmeyer, mentioned in Chapter 2, life consisted solely in motion. Organic substance was liquid and mobile in itself, which was why it was able to take on different forms in the first place.

Jean-Baptiste Lamarck (1744–1829), in his 1809 study *Philosophie zoologique*, famously made motion a fundamental principle of nature: "there continually reigns throughout the whole of nature a mighty activity, a succession of movements and transformations of all kinds." For all living bodies, their organization and "all their movements and modifications are entirely due to the movements of the various fluids occurring in these bodies," and "the fluids in question have by their movements organised these bodies."[31] Because the liquids flow constantly through the body, the "cellular tissue" that they form is also the solidified product of a movement — "the framework [*gangue*] in which all organisation has been built."[32] Within it, the organs take shape "by means of the movement of the contained fluids which gradually modifies the cellular tissue."[33]

In this vision of organic substance, the fluidity of organic matter was the basis of its plasticity and more specifically of the multifariousness and continual reshaping of organic life. Motion was the basic property of organic matter, and it was everywhere.

But motion was not the reason why an organism was thought to be alive: organisms came alive only through the organization of organic matter. Such organization, the result of motion, was tied in various ways to a vital principle, a life force, or vital properties, whether working from the outside or as a principle of the material itself. For Wolff, the motion of the nutritive fluids is caused by a *vis essentialis* that he does not define in any more detail; for Lamarck, it is an external "exciting cause," itself "purely physical," that "sets [the fluids] alone in motion" and thereby organizes the body.[34]

By the mid-nineteenth century, the microscopic study of motion had substantially transformed the notion of what it meant to be alive. For centuries, animals alone had epitomized life on the basis of their inherent motion; around 1800, when Lamarck and others were establishing the term "biology," it was organization and vital properties that assigned life by creating organic matter out of motion. With the advent of the cell theory, organisms came alive because their units, cells, were vital — vital in themselves. Now the cell, an enclosed unity that was not an animal, or organized, or moving, was declared alive.

This epistemic leap, which usually goes unnoticed, was anything but self-evident. It was a product of the cell's journey "from the microscope to the physiological laboratory."[35] On that journey, the cell became active, part and parcel of the physiological flow of the body. Movements streamed through it and all around it, constantly remaking it. The cell itself became less important than what it did. The container lost its appeal; the contents came alive.

Movement through Membranes

The Frenchman Henri Dutrochet (1776–1847) was key to thinking of the cell in physiological terms. Somewhat neglected today, or discussed only in relation to the studies of Theodor Schwann, Dutrochet's work is a major investigation into the physiological process of osmosis.[36] Dutrochet was proficient in medicine and natural history, a competent microscopist, and an admirer of Spallanzani's experimentalism. He was intrigued most of all by the most active organisms, whether the rotifers with their spinning wheels or touch-sensitive plants such as the mimosa.

Just as unhampered by disciplines or traditions as his scientific persona was his vision of cell physiology. All his life, Dutrochet had one overriding goal: to found a general physiology that would cover animals and plants alike. In 1824, in *Recherches anatomiques et physiologiques sur la structure intime des animaux et des végétaux,* he postulated that if physiology is to be established as a science, it must study everything that lives—"all living beings without exception."[37] Dutrochet complains that physiologists are still far removed from that goal, for everywhere there reigns an "isolation" between botany and zoology, pushing into the distance the "general science of life" that he finds so necessary.[38] In 1826, he pointedly opened *L'agent immédiat du mouvement vital* with the statement: "the science of life is a single science."[39] How does Dutrochet ground that science, and what is the nature of the life to which it will be dedicated?

Considered "in the physical order," writes Dutrochet, life "is nothing other than a movement."[40] A general physiology of plants and animals is thus a science of their motion, motion being, in Dutrochet's view, the crucial property of all living things without exception. In detail, Dutrochet distinguishes between different "faculties of movement," but all of them are rooted in a "single general faculty": "vital *motility*—this is life itself." All across the realms of animals and plants, this *motilité vitale* shows the same phenomena, only that among the plants these are "considerably less vigorous, considerably less developed."[41]

Dutrochet finds the empirical principle of movements within plants and animals in osmosis. Osmosis is the foundation of all physiological processes, the mechanism that gives rise to motion, nutrition, growth, development, and any other process in plants and animals.[42] The introduction to his 1826 study hence boldly announces that he has found nothing less than "the secret mechanism of vital movement."[43]

Dutrochet calls osmosis a "new physico-organic action."[44] He investigates its operation in the motion of cell sap in plants, distinguishing endosmosis (diffusion into the cell) from the reverse movement of exosmosis:

> When two liquids of different density or different chemical composition are separated by a thin and permeable partition, two currents establish themselves across that partition, led in opposite directions and unequal in strength. This results in the liquid accumulating more and more on the side toward which the stronger current is directed. These two currents exist in the hollow organs that make up organic tissues, and it is in this context that I have designated them "endosmosis" and "exosmosis."[45]

For concentrations of solutes in fluids to become equalized during osmosis, the key "vital faculty" is the plant cell's ability to become "turgid," that is, distended through the influx of fluid.[46] This is the default state of the cell: "vital movement consists in a necessary predominance of endosmosis over exosmosis, a predominance from which results the normal turgid state of the parts."[47]

This explanation places the cell at the core of physiological processes in the plant body (and also, as Dutrochet demonstrates, the body of the animal), since the "cells" or "vesiculae" and their membranes are the precondition for the process of osmosis to begin.[48] But Dutrochet's physiology depends upon *movements* within and through the cell. All plant or animal cells, he says, "are filled with organic substances" that are "sometimes liquid, sometimes thick, sometimes solid."[49] The cell "in its natural state, filled with an organic substance that is denser than water," carries out osmosis always "in relation to that liquid."[50] If one compares the cell's "extreme simplicity with the extreme diversity of its innermost nature," then its "uniformity of structure" only goes to prove that the diversity of physiological processes does not inhere in the cell itself, but stems from the differing nature of "the substances contained by the vesicular cells." This amazing "variety in the physical and chemical qualities of the substances secreted by the cells" is the hallmark of physiology.[51] In Dutrochet's view, the process of osmosis enables the cell even to exchange its contents completely: "the vesicles, the seat of a dual current of insertion and expulsion, tend by endosmosis to entirely renew their internal substance."[52]

Dutrochet builds his notion of osmosis out of the physiology of organic movement and only then identifies a general physical mechanism, which he can also provoke using "thin slices of inorganic bodies, permeable to liquids."[53] This is remarkable in two ways: Dutrochet reverses the usual sequence, that of seeking to demonstrate the effect of a physical mechanism in an organic process,[54] and he concludes that the phenomenon of osmosis is exclusively physiological not by its nature, but de facto — through the particular circumstances in which it occurs.[55] He makes the distinction that although there is a physical explanation of physiological motion, that explanation itself does not suffice to explain why, in nature, the phenomenon is actually found only in organisms. Whereas Dutrochet seems perfectly at ease with this subtle, yet significant difference, which underlines the limitations of physical explanations of organic processes, later biological research would ignore it entirely.

Discerning Motion

The physiology of cell motion raised an old question in a new form: How is it possible to discern movements at the level of the cell, movements of and within tissue?

For Dutrochet's study of plant sap, the difficulty was exacerbated by the "necessarily very slow" movement of the fluid.[56] Dutrochet was therefore surprised to read that his German colleague Carl Heinrich Schultz-Schultzenstein (1798–1871) had observed a rapid circulation of fluids. To understand the discrepancy, Dutrochet repeated Schultz-Schultzenstein's experiments. He first tested the optics of the microscope, using an instrument that had proved its worth in representing motion since the eighteenth century and was still being used in France, notably in François Magendie's lectures: the solar microscope.[57]

If the internal movement reported by Schultz-Schultzenstein "was an optical illusion," commented Dutrochet, when using the solar microscope "that illusion would have to disappear, since, in this kind of observation, the gaze is not directed to the object itself, but to its considerably enlarged image." In other words, if motion is seen on a screen, especially in magnification, then something becomes apparent that the object *itself* does not reveal. As it turns out, that does not amount to much: "everything seemed to me to be completely immobile."[58] To exclude the possibility of a simple experimental failure—the leaf examined might have been "burned by the concentration of the sun's rays"—he tries again, this time fixing the leaf on a glass slide with olive oil, but the result is unchanged. Dutrochet now accords primacy to the appearance produced by the solar microscope, judging the appearance of movement under the normal microscope to have been an optical illusion "produced by a certain play of light."[59]

Dutrochet, studying plant sap, was confronted with movements so slow that they could hardly be detected; Félix Dujardin, studying infusoria, was confronted with organisms almost too amorphous and translucent to be discerned. Hardly had one "recognized or discovered" some details, Dujardin writes, than "the memory of their forms" disappeared.[60] But Dujardin was well prepared for this difficult task. He had written his own handbook of microscopy and improved the microscopic visibility of translucent infusoria by inventing a special illumination apparatus that used lenses and prisms to redirect light onto the object and thus increase the microscope's magnifying power. An excellent draftsman, Dujardin had spent several months in 1819 as a student in the workshop of the French painter François Gérard (1770–1837), himself a pupil of the famous Jacques-Louis David; later, as a professor in Rennes, he turned his kitchen into an aquarium, with pots and pans as laboratory equipment for his infusions, and tried (albeit unsuccessfully) to put the new technique of daguerreotypy to use in his microscopical investigations.[61]

We have already encountered Dujardin's object of study, the amoeba, in the shape of Rösel's Proteus. Amoebae added further complexity to the issue of seeing movement, because all that can be seen of them is their motion. The amoeba displays, so to speak, movement stripped naked. Amoeboid movement is not mediated by any specific organs and is apparently devoid of organization—a granular flux, the peculiar movement achieved through the protrusion of cytoplasm to form pseudopodia.[62] The sequence of figures B1 to B7 in Dujardin's drawing (fig. 6.2) shows this kind of locomotion, where the substance inside the body first shifts in one direction and gathers, while in the other direction a slender protrusion takes shape. Through contraction, the granular substance is then squeezed into the protrusion. The process is repeated until, after around half an hour, the amoeba has reached the shape shown in B7 and dies.[63]

The amoeba consists of nothing more than a "gelatinous, semifluid mass that continuously, though slowly, takes on ever new forms."[64] This enables it to change its outer form completely in around two to four minutes. The organism takes thirty to forty minutes to cover the distance of a single millimeter, however, as its entire body mass seems to "flow in a certain direction over and over again," driven not by any cilium or filament, but only by an "inherent force."[65]

To represent the shape changing and motion that he has observed in gelatinous and transparent fluid, Dujardin reminds his readers, is to represent not the infusorians' "actual form," but only the effects of light differently reflected by the differently transparent parts of the organism. The forms he shows, thus, are not genuine outlines, but variables of the light: "results that vary according to the distance from the lens and the angle of incidence of the light passing the objects."[66]

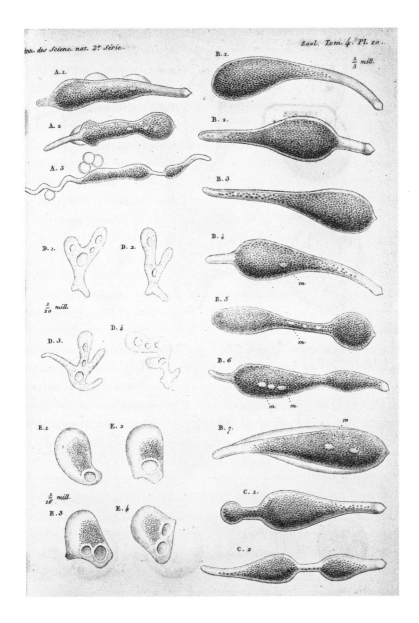

Figure 6.2. The development of motion in the amoeba's internal substance is shown in figs. B1 to B7. Dujardin, "Recherches sur les organismes inférieurs," 1835.

Dujardin calls the contractile, mobile substance he finds in the rhizopods and infusoria *sarcode*. This "glutinous, translucent substance, insoluble in water," which "contracts into globular accretions, sticks to the dissecting needle, and can be drawn out like mucus," is found in all the lower animals, "interspersed with the other structural elements."[67] A "perfectly homogeneous" substance or "living jelly," the mass is also "elastic and contractile" in itself, "extensible and able to spontaneously hollow out into spherical cavities or vacuoles."[68] Dujardin does not know how this contractility and the spontaneous formation of cavities comes about. One aspect of the sarcode is clear to him, however: "in a word, it is alive."[69]

Contractile Living Substance

Dujardin had observed contractility in infusoria, yet contractility was generally regarded as a property of higher animals—or more precisely, as we have seen, of animal muscles alone—and it required the stimulus of a nerve, if not the action of a commanding consciousness or soul.[70] The contractility found in completely unorganized organic material was therefore in urgent need of explanation. It raised fundamental questions: Why are infusoria able to contract at all? And, even more importantly, if movement can be found in unicellular infusoria and the boundaries between the lower plants and animals are blurred, does that mean there is only one living substance, identical in all of organic nature?

The answer was far from obvious, since the fact that cells were the structural building blocks of both plants and animals did not automatically imply that their organic substance was also identical. Neither was there agreement as to the precise nature of the contractile movement. Carl Theodor von Siebold, for example, despite being the first to identify the infusoria as unicellular organisms and recognize the congruence of plants and animals in their cellular construction, made the peculiar motion of the infusoria the criterion for defining them as animals. Siebold distinguishes the "voluntary" movements of the infusoria from the involuntary ones of the lower plants and associates "free expressions of will"—that is, "the capacity for voluntary contraction and expansion of the simple, cellular body"—with the animal. This he contrasts with the "local movements" of the plant, which are not induced by a voluntarily contractable and expandable "parenchyma."[71] Contractility here seems to be a property not of vegetable, but of animal tissue, and a form of expression of will that is exclusive to animals, even if those animals are mere infusoria.

Whereas Siebold believed this distinction between animals and plants had been unjustly neglected, others were beginning to question its usefulness. For Siebold's contemporary, the German zoologist Alexander Ecker (1816–1887), investigating contractility was a comparative matter—indeed, "one of the most important tasks of a comparative histology and physiology."[72] Ecker addressed himself to the "extraordinary contractility" of the polyp Hydra, finding the same properties that Dujardin had discovered in the sarcode of the infusoria: the polyp consists of a "uniform, partly clear, partly granular, soft, distensible, elastic, and contractile substance that is perforated like a net, with a more or less clear fluid contained in the cavities."[73] Due to this inherent contractility of the material, which may in more highly developed animals form muscle mass, Ecker chooses the description "unformed contractile substance."[74] The critical point is that contractility is not the result of an organization within the muscle, but is inherent to the material itself. Only more detailed study of the chemical constitution of the organic material, Ecker notes, will reveal the cause of its contractility and clarify whether contractility is a property of "very different substances" or "is tied to a particular chemical constitution in changing histological compositions."[75]

In the botany of the same period, Hugo von Mohl (1805–1872) used the term *protoplasm* in 1846 to describe plants' "viscous, colorless mass, mixed with fine granules" and not enclosed by a membrane.[76] He chose the name to highlight the physiological function he attributed to the protoplasm, that of supplying the material for the formation of further cell structures in the plant.[77] The botanist

Franz Unger (1800–1870) also credited plant substance with contractility and characterized the alternation of contraction and expansion as "progressing rhythmically." Unger described the contractile substance of the protoplasm in detail, explaining that it "sends forth parts of its colorless body in every direction in the form of threadlike extensions that, initially broad, very soon divide and ramify." Chemically, he defined the protoplasm as a protein substance and a "mainly nitrogenous body."[78]

This "material [*Stoff*] of the plant cell," Ferdinand Cohn (1828–1898) agreed, was the "principal seat of almost all the vital activities, but especially all the phenomena of motion inside the plant cell," and showed the same qualities as the microorganisms described by Dujardin and Ecker.[79] For the important group of the slime molds (today no longer classified as fungi), the botanist and mycologist Anton de Bary (1831–1888), too, attributed to the plant protoplasm "the capacity for independent contraction, movement, and shape changing, not derivable from any external cause, to an often very high degree"; in addition, protoplasm was very similar to sarcode in its chemical composition.[80]

In 1850, after studying all sorts of life — plants, infusoria, and algae — Cohn was the first to conclude, "with all the certainty that can attend any empirical deduction in this field, that the protoplasm of the botanists and the contractile substance and sarcode of the zoologists must be, if not identical, then to a degree analogous formations."[81]

The identity of sarcode and protoplasm meant that whether plant or animal, the organic substance shows the same behavior in optical, chemical, and physical terms.[82] It also meant that contractility would henceforth be regarded as a central property of all organic material, including that of the plant. Its concrete expression, according to Cohn, is "the capacity to be induced by external stimuli to change form temporarily," and it is, "like assimilation, respiration, fluid conduction, reproduction, etc., a vital activity of the cell as such."[83]

Materially, then, the protoplasm of animal and plant seemed to be largely identical. But Max Schultze (1825–1874) soon went further than that: denying that the cell membrane is a necessary component of the cell, he declared the cell to be merely a "little lump of protoplasm with a nucleus and no envelope," a "spherical clump" that is "held together by its own consistency."[84] At bottom, what makes up both animal and plant is nothing more and nothing less than a substance with the as yet unexplained property of intrinsic mobility.[85] For this "inner cause" of movements, wrote Schultze, contemporary science had no other term available than "contractility."[86]

Matter Comes Alive

Recognizing the identity of the organic substance in both animals and plants and declaring cells to be sheer organic substance did not make the substance's intrinsic motion any the less puzzling. On the contrary, once protoplasm was freed of the boundaries of the cell, no longer confined to its container, its movements became even more ubiquitous:

> Just as unenclosed cells, consisting only of protoplasm and nucleus, in the course of their life quite ordinarily execute amoebalike movements even if they are components of another organism, or just as the protoplasm inside rigid-walled cells, where external changes of shape are no longer possible, can execute highly complicated, independent movements; so we should not be surprised if these movements occur even more vigorously in cells that themselves constitute a whole organism, either alone or in combination with just a few others.[87]

When Ferdinand Cohn saw these movements for the first time, he found them both "marvelous" and "alarming for any enemy of mysticism, because they have no perceptible cause."[88] In 1863, two centuries after Leeuwenhoek, but in something very close to his words, Max Schultze describes the movements as "constantly recaptivating the eye," "a surprising and truly magnificent spectacle" in which the protoplasm, like a group of walkers strolling along a broad avenue, sometimes "faltering and trembling," "suddenly stops" and begins to move again "only as if after a short pause for thought." "I believe," writes Schultze, "that we are still very far from having a sufficient overview of the processes of organic movements even in their simplest manifestation."[89]

If protoplasm is itself alive, that is because it moves. It is the activity of organic substance, and the activity alone, that indicates lifelike qualities. But assuming that life is movement in the interior of plants, animals, and microorganisms—that protoplasm is contractile and flows of movement incessantly make their way through cells, tissue, and organisms—how do the vital processes arise out of these movements? What are nutrition, growth, and development from the perspective of motion? How can they arise from motion, function through motion, be perpetuated in motion? Finally, how does motion generate form?

CHAPTER SEVEN

The Study of Formation

Morphogenesis is the process by which the organism develops its shape. Arguably, this becoming of form is the fundamental process in nature and in a living being.

In the first decades of the twentieth century, two German scientists, the embryologist Ludwig Gräper (1882–1937) in Jena and the Berlin-based anatomist and embryologist Friedrich Kopsch (1868–1955), approached the problem of organic formation from the novel perspective of cell movements. In seminal work that is largely forgotten today, they turned their attention to the study of gastrulation, one of the pivotal phases in early embryogenesis.

Since the end of the nineteenth century, it had become apparent that cell movements played a highly significant part in the gradual shaping of the embryo's vessels, tissues, and organs. At the same time, making sense of cells in motion faced scientists with a troublesome quandary. For cells could be observed in motion in the living specimen — but the organisms examined under the microscope were stained, fixed, and sliced. In these dead substances, movement could not be seen, so how could it be thought? Conversely, when studied in the living specimen and thus visible, but ephemeral, how could movement be captured, traced, and archived? Once that problem was solved, that is, once cells became visible and could be investigated as moving actors, a further, serious problem arose:

How could the emergence of form be interpreted out of hundreds of diverse, delicate, and seemingly erratic movements?

Despite their neglected status in current embryological research, Kopsch and Gräper were at the forefront of the science of their day. Embryology was one of the first fields in biology to use the novel technology of cinematography, and gastrulation was one of the earliest embryological problems to be researched using film, as Kopsch and Gräper did. The course of Ludwig Gräper's academic career, especially, offers a rare glimpse into the making of a research program solely devoted to the study of cells in motion. In terms of the material history of motion research, moreover, Gräper's is among the few cases in which cinematographic material from the 1930s has survived to this day.

But the journey toward seeing cells move was a long and difficult one, and what might at first seem like a perfect solution to the problem of investigating motion—capturing movement in moving images—did not in fact solve the enigma of cell movements in gastrulation. As we will see, Gräper ultimately took a different path and found that cells are matter that dances.[1]

Seeing Movement, Understanding Form
Looking at the history of morphogenesis from the point of view of cells in motion brings us back to the question of how to experiment on a moving world. Whether dead in a drawing or staged on a white wall in all their vitality, insects and infusoria traveled between image and projection in the work of Leeuwenhoek and Rösel and between representation and imagination in the mind of the viewer. The amoeba was too mutable to remain fixed even in memory. It presented the observer with the almost insurmountable task of seeing form where there was no form to see and thus science with the question of whether something was an animal that could not by rights *be* an animal, seeming as it did nothing more than a formless mass.

Starting in the mid-eighteenth century, epigenetic theories of development found themselves facing the same conundrum: how to represent something that could be comprehended only in that very act of representation. For centuries, the coming into being of new life had been an act of divine creation and the development of new forms merely the growth of existing ones. Epigenesis, instead, argued that form takes shape gradually out of homogeneous matter. In the 1750s, Caspar Friedrich Wolff struggled hard to discern the alternating play of flow and standstill when he observed the nutritive fluid moving in plant and animal substance; in the 1830s, Christian Heinrich Pander and Karl Ernst von Baer described the gradual formation of all the organs in the developing embryo based on the concept of a highly orchestrated spatiotemporal process of membrane folding.

A new approach in epistemological terms, the epigenetic theory of generation also implemented a new visual stance. Because understanding formation now depended on observing the gradual establishment of structures in the developing embryo, the gradual coming-to-be and continuous changing of form, visual strategies to depict these processes and their temporal patterns were crucial. Image series, especially, were at once a new representational resource to capture change and a new epistemological operation to construct it. Series of images enabled the relationship of forms — or rather forms *as* relationships — to emerge visually.[2]

When scientists at the end of the nineteenth century turned their attention to gastrulation, they had to grapple once again with the ongoing cognitive and visual challenge of seeing movement and understanding form, or, put the other way around, of seeing form and understanding how it is created through motion.

Gastrulation is the most dramatic and puzzling of the morphogenetic events that occur in the early embryo. Following the formation of the blastula (a hollow sphere made of cells called blastomeres), the blastula's single layer differentiates into the three germ layers, the ectoderm, mesoderm, and entoderm, through various forms of inward folding, curving, and bending. Gastrulation differs considerably between different groups of animals; in birds, the organism studied by Gräper, it begins with the formation of the primitive streak. Generally speaking, the process draws inside the embryo cells that lie further outside in order to form the germ layers, which will give rise to all the animal's future structures and organs. Gastrulation, then, marks a period of major cell movements, translocations, and rearrangements with tremendous activity inside the ovum. Its common characteristics are changes in cell motility, cell adhesion, and cell shape, but the exact ways in which the cells move are manifold, showing great spatiotemporal variability and complexity both within and across species.

Ludwig Gräper and his research on gastrulation bring us to a problem pivotal to the study of movement in the microscopical world: in morphogenesis, we cannot easily discern either form or movement. This is not only a problem of seeing and interpreting, of making perceptible and intelligible at the same time. Rather, movement and shape, behavior and form have to be figured out in relation to one another.

Fixed in Paraffin, Fixed in Time

Before he went to Jena, Gräper studied at Leipzig University with Carl Rabl, who had followed Wilhelm His as head of the anatomy department there in 1904.[3] Gräper took an interest in movement right from the start of his career, while still a student with Rabl. In his 1907 paper "Untersuchungen über die Herzbildung der Vögel" (Studies on the formation of the avian heart), he investigated germinal disks to understand early embryological formation in birds.

As Gräper notes there, among the very first structures to develop in the ovum is the vascular system. More precisely, the vascular system begins to take shape as small islands of blood. These islands intrigue Gräper as "probably still the most obscure, but also the most interesting point in the whole developmental history of the vascular system."[4] In the germinal disks he has at hand, five altogether, he finds "entirely isolated blood islands," sometimes "in substantial numbers and at a great distance from the mesoderm and from other blood islands."[5]

Gräper's most important question with regard to the formation of the islands is whether the cells arise afresh (through cell division) or migrate from other parts of the ovum, and if the latter, from which of the germ layers, the mesoderm or the entoderm. His interest, then, was movement, but his object of study was the dead, stained specimen and his method a classic histological investigation.

The technique of fixing and staining microscopy specimens has a long history in the study of formation. Albrecht von Haller had stained chick embryos as early as 1738; in the late eighteenth and mid-nineteenth centuries, carmine and rose madder were used as dyes. At the beginning of the twentieth century, Gräper was subjecting his embryos to a complex and time-consuming histological procedure. They were fixed and hardened using a mixture of picric acid and mercuric chloride, stained with alcoholic borax carmine, embedded in paraffin, cut into slices just 7.5 μ thick, attached to the slide with a mixture of clove oil and collodion, and finally stained blue and red again using hemalum and erythrosine.[6]

For Gräper in this phase of his career, studying motion means measuring distances: he measures the intervals between individual cells or cell agglomerations in the blood islands and the germ layers and from his comparative measurements infers that the cells migrate within the ovum, moving from the innermost germ layer, the entoderm, out to the blood islands. If the cells had come from the mesoderm, they would have had to cover a very "long distance, up to 3 mm"—a route too long and arduous to be worth considering. Because understanding motion "depends on ascertaining the connections between cells," yet Gräper can study only fixed material, he has no other option than meticulous measurement of distances and the intervals between individual cells and cell agglomerations.[7]

Despite measuring more germinal disks and additional bird species (seagulls and ducks) in the years that followed, Gräper did not manage to determine the necessary "fixed measuring point" in the embryo with the ruler alone and was therefore unable to reach "absolutely binding conclusions regarding the movements of growth."[8] This prompted him to modify his method and turn to a new medium.

The Temporary Preservation of Life

The new method Gräper adopted was vital staining. Despite its name, "vital staining" did not mean that Gräper studied living embryos. Although what was stained was a living organism, the embryo did not so much continue to live as merely fail to die for one or two days longer. Gräper had started to experiment with the dye neutral red as early as 1905. The art of staining, he wrote in 1911, consists in retarding death for as long as possible, if necessary "until shortly before hatching."[9] Though "hardly gentle" to the organic substance, coloring the tissue "an intense pale yellow" and "a dark brownish red," vital staining did have the critical advantage of opening the path to a new technique: serial photography.[10]

Gräper explains that photography enables him to "make visible embryos that have hitherto been as good as invisible."[11] It is also "far superior to observation by microscope and micrometer in evaluating the processes of growth." But photography's real advantage does not lie in its representational quality, its ability to give greater precision and a more detailed recording of what has been seen. Far from it: Gräper emphasizes that "one naturally sees much more with the binocular microscope than on the photographic plate." What photography can offer, rather, is "the repeated comparison of the images, not only in a straight sequence, but also in reverse order." The primary benefit of serial photography, then, is to measure, and more specifically to measure comparatively. The photographs make it possible to compare serially—and to do so on the basis of a series of different embryos, since one cannot "see all the movement processes in the photo series of a single embryo."[12]

The two plates in Figure 7.1 show a selection of sixteen photographs taken from the series of one embryo, "U." The series enables Gräper to track the motion of individual structures, such as Hensen's node, a regional thickening of cells at the anterior end

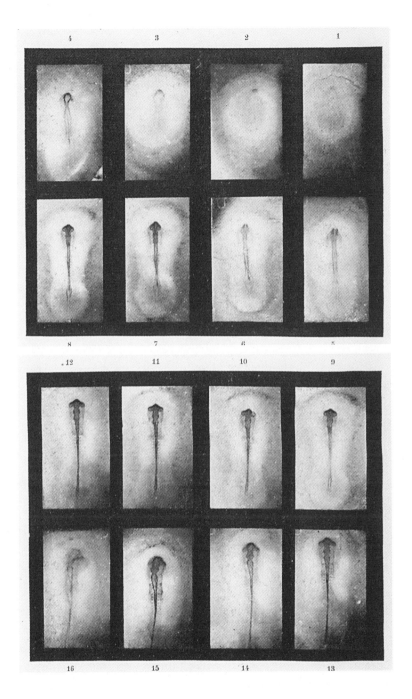

Figure 7.1. Serial photographs taken during the development of a single embryo. Gräper, "Beobachtung von Wachstumsvorgängen," 1911.

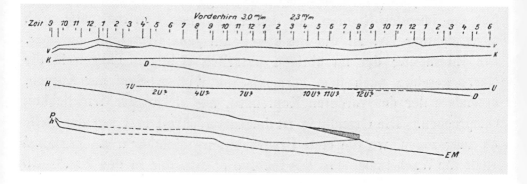

Figure 7.2. Gräper's diagram shows the lines of movement taken from individual measurements of different structures. Gräper, "Beobachtung von Wachstumsvorgängen," 1911.

of the primitive streak, which in Gräper's figure 1 tips the primitive streak and in figures 3 and 4 migrates backward. Gräper describes this backward movement of Hensen's node as a wave or vertical motion in which the cells do not move across the surface, but ever new cells move up from deeper in the embryo's interior and form the node.[13]

Aggregating measurements based upon these photographs, Gräper is able to compile the growth curve of an embryo (fig. 7.2). Following the lines in the diagram from left to right, we can track the movement of different structures in the embryo *U* over a period of thirty-three hours. From top to bottom, the diagram shows a cross section of the ovum, with individual letters designating different structures: from *v*, the anterior end of the area pellucida, through *H*, Hensen's node, and *P*, the posterior end of the primitive groove, to *h*, the posterior end of the area pellucida.[14]

The diagram orients in time and space the shifts that occur during development. Dashed lines track structures that were sometimes out of focus and thus show partially ambiguous lines of motion; an almost entirely smooth line such as that of *H* marks an "always constant speed" of movement, its downward direction in the diagram indicating the inward turn of the node in the egg's interior. The crosshatching at the end of the motion of Hensen's node, in contrast, shows that here Gräper was not able to determine precisely in space and time the events that occur when the node meets the posterior end of the primitive groove and the latter "flattens out from the back to the front."[15]

It is therefore not the photograph, but the line of movement, as the substrate of his measurements, that permits Gräper to analyze motion. He extracts motion out of the photograph in order to capture it analytically in the precision of a graph (and where that endeavor fails, indicates as much using crosshatching or dashed lines). Gräper's contemporary Friedrich Kopsch, working with living specimens, uses photography to the reverse effect. In Kopsch's photographs, motion is nothing but a blurred intimation in the image. This is exactly the added value that only photography can supply.

Exposing the Living

Kopsch was one of the pioneers of time-lapse serial photography, and in fact the first to provide concrete evidence of cell movement in gastrulation.[16] After training under Oskar Hertwig and Wilhelm Waldeyer, he held a teaching position as prosector for the anatomy department at the University of Berlin for many years, though he became a full professor of anatomy and embryology there only in 1935. His anatomical knowledge fed into the many editions of his handbook on human anatomy, for which he is mostly known today.[17]

As early as 1895, Kopsch conceived of gastrulation as an event of "cell displacement."[18] Yet investigating the movement of cells was by no means a matter of course, and it prompted major methodological reflection on his part. For Kopsch, it was essential to keep the embryo alive and trace cell movement and behavior continuously in a single living specimen, avoiding artificial conditions to the greatest possible extent.[19]

He concluded that working with histological sections, the usual approach at the time, yielded findings that were too "diverse" and "subjective" to be valuable. It was of prime importance to follow the displacement of cells "in one and the same egg."[20] Unlike Gräper, Kopsch studied not birds, but amphibians. Due to the amphibian egg's size, opacity, and spherical shape, only movements on the surface of living eggs are discernible, he found. In addition, cell movements in gastrulation take place in the vegetal hemisphere of the egg, facing away from the observer.[21] To enable observations, the microscope has to be inverted.[22] No less experimentally challenging, to Kopsch's mind, is the need to ensure the visibility of each individual cell. Kopsch rejects the placing of experimental or natural marks and fixing the cells in a set position, as practiced by his contemporaries Wilhelm Roux, Oskar Schultze, and Eduard Pflüger. This is because it is impossible to know how much destruction in the cells is caused by these methods; if the amount of dead cellular material is large, it might modify the regular course of gastrular movements.[23]

Studying live axolotl and frogs, Kopsch made continuous series of microphotographs, taken at intervals of one to several hours. But his aims in employing serial photography were different from those of Gräper, and so were his photographic methods. He exposed his photographs for twenty to thirty minutes, an exceptionally long exposure time — he terms his photographs "continued shots" — that inscribes mobile cells as blurred traces onto the paper, their nuclei taking the shape of strokes rather than spots, whereas immobile cells are clearly discernible by their sharp cell borders.[24]

To be sure, Kopsch also uses his photographs for measurement. These measurements, however, translate motion not into a diagram, but into a calculation. Kopsch's goal is to work out the distance traveled by the cells over the course of gastrulation. To that end, in the individual photographs of the axolotl egg, he measures the distance between the cell nuclei, arriving at an average distance of 0.65 mm. The cell nuclei exchange places within the thirty-minute exposure time of the photograph, so that in this half hour, the cells themselves must travel just that distance. Because the cells do not move at an identical rate in all regions of the ovum, Kopsch calculates a mean speed, from which he extrapolates the total duration of gastrulation in the axolotl to be thirty-six hours. Based on these calculations, he concludes that all the cells that were close to the blastoral lip at the beginning of gastrulation have arrived at the edge of the blastopore by the time gastrulation comes to an end.[25]

Moving Cells, Moving Images

Photographs and series, lines and numbers, aggregated diagrams and calculated distances were all indicators of movement, but they could not compete with actual observation when it came to evoking the living reality of motion, let alone full extent of the bustling activity inside the ovum. A decade after his measurements and photographs of 1911, therefore, Gräper moved on. He tailored the latest innovations in cinematography to his research in the hope of finally being able to capture and represent motion as actual movement.

In 1926, he showed his colleagues at the Anatomical Society in Freiburg what he called "cinema images" of the early development of avian embryos.[26] His report on this research concedes that "one may wonder why I devote so much effort to a rather technical matter," yet, clearly, far more is at stake than a mere technicality. The advantage of the "running image," Gräper explains, is that it permits the developmental movements to be "recognized immediately" and with such "convincing clarity" that a single "improvised cinematic projection" is more persuasive than a hundred carefully measured germinal disks. The projections are especially useful for the less dedicated student, who is unlikely to forget the images he has seen; for the experimenter, their value lies in the "intimate glimpse" that they offer into "the morphology and physiology of development, the operation of color, light, temperature, and all kinds of damage."[27] Film also gives the researcher the opportunity to "truly 'take part' in the events."[28]

Gräper was not the first scientist to make use of cinematography.

Neither was he the first to accentuate its remarkably immediate, vivid, and persuasive character—these advantages of film had been stressed many times since the early cinematography of Julius Ries, Louise Chevreton, Frederic Vlès, and Jean Comandon at the turn of the century. What attracted Gräper more than film's technical affordances and vivid "descriptiveness" was its potential as a form of research, one superior to experimentation.[29]

Of all scientific methods, Gräper contends, only continuous filming of a single living specimen can do justice to the enormous variability and individuality that characterizes developmental processes. To distinguish normal variability from pathological deviation, especially, continued filming throughout the whole of embryogenesis is more accurate than experimentation.[30] Most importantly, research on large numbers of fixed specimens results only in statistical values, and sections of embryos offer no more than a "dense succession of snapshots" that is insufficient to determine exactly "which points of a previous stage correspond to particular succeeding ones" or to track actual movements in living cells. Accordingly, Gräper judged that his "precisely measured serial photographs" of 1911 had permitted only "isolated conclusions" to be drawn, and he anticipated "special progress" through "the cinematographic, time-lapse representation of the living embryo."[31]

When he turned to film in the 1920s, Gräper was entering the last decade of his career. He finally started filming when the German film company Ufa inquired about his work with film and he began to collaborate with the optical devices manufacturer Carl Zeiss of Jena, notably with the Zeiss-based physicist Henry Siedentopf (1872–1940). A pioneering microscopist and the head of Zeiss's microscopy department from 1907 to 1937, Siedentopf worked on rapidly advancing new technologies such as microphotography, time-lapse, and microcinematography.[32] Gräper hoped that cinematographic techniques would enable scientific disputes not only to be approached in a new way, but to be "settled" once and for all.[33] For twenty years, he had been working on the question that had occupied him from the very start: What is the source of the cells that give rise to the first structures in the avian egg?

Unfortunately, Gräper's 1926 film on the chick embryo has not survived, and he never published on it in any detail.[34] But as he explained to his colleagues at the screening, the film had indeed allowed him to resolve one highly contested question: it showed that the vessels in the embryo arise not through cell division, but through the displacement of cell material—that is, they are formed by the migration of cells that originate in the area opaca and move into the area pellucida.[35]

Yet cell movements and migrations prove to be even more multiple, locally diverse, and varied in their spatial orientations. The motion that Gräper observes in one region of the ovum is accompanied by a large number of other displacements in other regions. While the primitive streak is "growing backward," Gräper can also discern a "lively backward flow of growth" at the side of the area pellucida's posterior end. Then the flow can "suddenly" be seen to turn back and begin to move first forward, then sideways. He describes how, in order to form the medullary folds, material is pushed from the area opaca forward and sideways in an "arching" movement, and the cells forming the amnion are transported "both from behind and in a twirling movement from the front."[36]

The thorny question of cell movements was therefore far from being settled; the countermotions and shifts of direction revealed by the film only made it more complicated. Cinematography was also unable to resolve a central problem: Gräper observed that the movements during embryogenesis took place in different tissue layers, but cinematic images cannot accurately identify the depth of tissue at which the various movements lead to the primitive streak's formation. Gräper himself remarked on this dilemma:

> In my early cinematic recordings of living chick embryos, in the middle and at the posterior end of the germinal disk I had discerned currents that sometimes seemed to crosscut each other. From the start, it seemed probable to me that this was an illusion, and that if I were to succeed in distinguishing

the currents according to their depth, they would show a lawful or at least regular behavior that would offer scientifically valuable insights into processes during the formation of the germ layers and the gastrulation of birds — processes about which very little is yet known, but that are all the more shrouded in theories.[37]

Film showed motion with vivid immediacy in a live setting, but did not in fact explain formation. In order to understand how the manifold cell displacements observed in gastrulation contribute to the formation of organic structures, another step was required, one that no longer simply visualized movement as movement, but made it into an object of scientific analysis. That step was stereo-cinematography. Gräper was the first to build and deploy a stereocinematographic apparatus,[38] and it finally enabled him to solve the mystery of how cells move in opposite directions and different dimensions within the chick embryo. Stereocinematography revealed that they are dancing.

Spatial Depth and Temporal Displacement

At first sight, bringing to light the dance of the cells might seem a purely technical challenge. We need spatial depth and differentiation if we are to discover where the cells come from and exactly how they are exchanged between the surface and interior of the egg to form the primitive streak in the center; stereocinematography offered Gräper the high degree of three-dimensionality required to see and depict those movements.

In stereoscopy (to put it rather simply), two images are recorded separately for the left and right eye, and when they are presented with the help of special glasses, mirrors, or prisms, the perception of a single image with a particular spatial depth is created.[39] The stereocinematographic apparatus that Gräper built at the anatomy department in Jena, though, was far more than an optical device. Its numerous components included a recording device with the stereoscopic additions, such as prisms and mirrors, necessary to enhance three-dimensionality to less than a tenth of a millimeter and to enable the images to be turned upside down as often as required; the objective and microscope, housed in a heating cabinet; the time-lapse apparatus; the lighting apparatus; and the negative film. Gräper experimented with film from different manufacturers, made of different materials and with differing sensitivity to light. Central to the setup, of course, was the vital specimen itself, the stained embryo, which had to be carefully prepared and kept alive in the incubator at least for the duration of filming.[40]

The result was a film that showed upright, stereoscopic "dual images," as Gräper called them. It could be played back with a normal projection apparatus, although the images then had to be juxtaposed "using two prisms set before the eyes at right angles, suitably connected to each other," in such a way as to produce a three-dimensional impression. Because the film was laterally reversed when recorded, it needed to be fed into the device the other way around, and the apparatus had to "run backward" for the process to appear in the correct temporal sequence.[41]

This technical complexity meant that stereoscopic projection was a rare event. In fact, as Gräper explains in his report, the impression of depth achieved in the projection is anyway insufficient, which is why he constructed a special stereo magnifying glass in order to look at the film directly. It was only by "diligently working through" individual recordings that he was able to perceive the necessary depth relationships.[42]

Figure 7.3 illustrates how Gräper set stereoscopy to work in order to understand cell movements. His figure 8a shows an embryo thirty minutes after the start of filming, which lasts for a total of nine hours, and seventeen and a half hours after the beginning of incubation. Figures 8b and 8c both show the embryo a further eighteen minutes later. When, as here, the images are recorded with a time lag, the stereo recording makes it possible to view nonsimultaneous events simultaneously, which stereoscopic viewing shows as a movement in space: looking at images a and b stereoscopically,

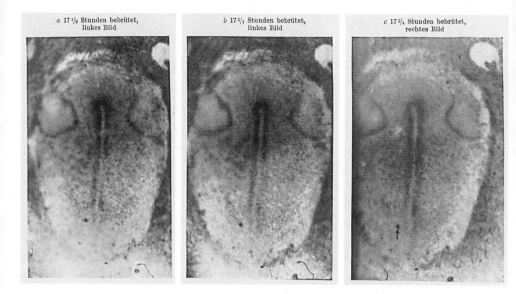

Figure 7.3. Gräper devotes a total of fifty stereocinematographic images to the polonaise. The arrow in the lower part of image "c" of the series shows the displacement of cellular material in the depths of the embryo. Gräper, "Die Primitiventwicklung des Hühnchens nach stereokinematographischen Untersuchungen," 1929.

movements on the surface can be seen in a shift to the left, movements in deeper layers in a shift to the right. If images b and c, recorded simultaneously, are combined stereoscopically, they give an impression of the depth.[43]

Polonaise

Stereocinematography was more than a technique for creating a novel perception of physical space — it also opened up new epistemological spaces. Choosing and juxtaposing images taken simultaneously or with a slight delay allowed Gräper to switch freely between seeing deep into the tissue and following cells as they move in time. He was able to play with space and motion, to use one in order to illuminate the other, and to exploit images to the maximum by relating them to time and space in multiple configurations.

Through the resources of stereoscopy, Gräper's investigation of chick embryos revealed that the primitive streak forms through a "double countercurrent," an elaborate, highly ordered, and intricately patterned cell movement. To make sense of the movement, Gräper compared it to a dance — specifically, the polonaise: "The couples move along in two columns on opposite sides of the ballroom, swing in toward the middle at the end and, joining hands, move forward through the center in fours."[44]

Just like couples on the ballroom floor, cells dance inside the ovum. The caudal cell masses "swing in toward the middle" and move upward at the same time as the cranial and lateral masses move toward the tail end.[45] In additional movements, cell material is exchanged between surface and interior, as "forward-flowing material" in the layer closest to the surface is replaced by other material that flows from the center and the sides.[46]

Gräper uses the figure of dance to grasp the sequence of movements holistically. A dance is a closed, yet mobile formation. It is defined by the relationships of its constituent parts, constant as a pattern, but not constant in the individual parameters of its performance — such as when, how fast, how often, at what points, or in what precise extension the dancers move through space. The dance is a choreography or pattern of movements that sustains itself despite variable conditions. Unlike his measurements and diagrams, all of which presented isolated or aggregated data, the imagery of dance enabled Gräper to apprehend the entirety of the movement, the interplay of the cells as a single orchestration in time and space.

Finding the cells' dance resolved two of gastrulation's great puzzles. One is the complex movement of double contrary flows, or polonaise, that is described by spatial differentiation during gastrulation in the individual organism. The other, far more difficult one is the great variability shown by organic development as a whole.

Individual embryos grow and develop at very different rates, notes Gräper, but regardless of their "very variable appearance," they can all become normal embryos.[47] This is why the continuous recording of the developmental process, possible only by means of cinematography, is so crucial. Gräper remarked many times that neither a series of single-state images—however smooth and closely spaced the series—nor "statistically calculated measurements" or experimental markings offered the dependability that he needed in order to define the successive stages of development precisely.[48]

The variability in individual ontogenesis is also to be found in the gastrulation of different species. Science, writes Gräper, has long grappled with the "unusual difficulties" posed by the "extraordinarily divergent pictures of development" that different vertebrates display. Gräper finds the reason for the diversity of development in different animal species in the "time lags" of developmental processes. The early developmental events are all "the same in principle," but their often considerable divergences result from the temporal variation in the sequences by which they proceed.[49]

The "complete, simultaneous impression of depth and motion,"[50] the contemplation of movements in their totality as a coordinated dance, is what tells Gräper that these are not distinct forms of gastrulation. Instead, the spatiotemporal choreography makes what is fundamentally the same event into many, highly differentiated gastrulation processes. In the bird, for example, the polonaise and the invagination of the mesoderm are "two completely different processes that succeed each other in time," whereas in the amphibian, "the two processes occur in a single, jointly resulting movement."[51]

Movement (whether collective or of single cells) and development, then, vary greatly both between different species and within a single species: the same structures might be the result of many different chains of events. Movements can be extremely diverse, yet still result in the formation of the same structures. Early development in all vertebrates, on this view, has "the same starting point and the same goal"; only the intervening route differs.[52] By using the image of the polonaise, Gräper was able to describe just that route.

Research and Rhetoric

The dance is not revealed by simply looking. Kopsch's and Gräper's diagrams and calculations; their photographs (whether long exposures, continuous series, or separately for the left and right eye); their making-visible and experimental manipulations; the calibration of the living against the dead and vice versa—all these techniques, preparations, and forms of representation had to be applied in concert if patterned, coordinated movements were to be intellectually and visually grasped.

As such, film in isolation proved not to be quite the congenial research method that Gräper had initially imagined. It was unimpeachable in one respect, however. In 1926, Gräper had particularly stressed the memorable and persuasive character of film for students, and in the late 1930s, when he returned to cinematography, he made three films—all still extant—at the request of the Reich Office for Educational Film (RfdU). The RfdU was created in 1934 as part of the Reich Ministry for Science, Education, and Culture, and from 1935 on had a special section devoted to universities.[53] The production of films in collaboration with the RfdU necessarily entailed a didactic perspective. In fact, a persuasive demonstration of his findings was exactly what Gräper sought in the late 1930s—though his reasons were less didactic than rhetorical.

When, in the mid-1930s, several scientists cast doubt on his discovery of the polonaise movement, Gräper recalls, he found himself "compelled to demonstrate the movement in a way so evident to the eye that even my most vehement opponent must be convinced."[54] The results are his three last films, made in 1936 and 1937. Gräper having died during the work, the final versions were completed by his assistant Irmgard Weiberlenn and the accompanying booklets written by his student Hildegard Treiber-Merbach.

One of the films is dedicated to gastrulation and the formation of the primitive streak in the chick embryo.[55] The film is composed of a selection of scenes. An intertitle first announces what the viewer will be able to observe; white sketches on a black

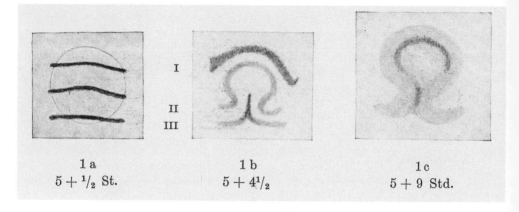

Figure 7.4. The rearrangement of the cell material during gastrulation can be tracked along the deformation of a straight line into an omega shape. The sketches, published in Gräper's article "Die Gastrulation nach Zeitaufnahmen linear markierter Hühnerkeime" (1937), are not identical with any of those shown in the film. The closest to them are the sketch and subsequent films with three color markers starting at minute 4:03.

background follow, drawings that explain schematically what is then seen in the film sequence, played back at around two thousand times the speed of the natural process. The polonaise flux of the cell movements is made visible by color marking with the dye Nile blue sulfate. A few hours after the beginning of incubation, color marking is applied in the young germ disks in straight lines across what will later be the longitudinal axis of the embryo (fig. 7.4). The changes in the course of these marked lines reveal the progress of the polonaise. More precisely, the process becomes visible in the bending of a straight horizontal line into the shape of the Greek letter omega. From this transformation, we can read the displacement of laterally located cell material backward and centrally and the displacement of centrally located cell material forward.[56]

Gräper had already made use of color marking in his stereo-cinematographic experiments during the late 1920s. At that time, it served him exclusively as a control, a way of supplementing the stereo film. Color marking on its own, he insisted, could "never with certainty" supply insight into the movements of the cell flows.[57] In the late 1930s, however, Gräper did not hesitate to demonstrate with film something that he had not been able to discover with it: the polonaise.

Its impressive visual rhetoric notwithstanding, film did not manage to draw cell movement into the limelight of embryological research in the twentieth century. As well as Kopsch and Gräper, G. Frommolt of the University of Berlin's medical cinematography department and Willi Kuhl and Hans Freska of the cinematic cell research institute at Frankfurt University used film for embryological studies in the period immediately before and during the Second World War, but the war called a halt to the groundbreaking work in cell movement and cinematography being done in Germany. A generation of embryologists had died (Gräper in 1937) or else returned only briefly to research, as Kopsch did when the University of Berlin reopened in 1946.[58]

Motion nonetheless made its presence felt in twentieth-century embryology. Particularly in the anglophone world, a culture of research using film was established in the 1920s and sustained across the caesura of the Second World War by a number of British and American scientists and institutions.[59] And at the sites of work on film, the questions raised by Kopsch and Gräper had lost none of their relevance half a century on.

In the 1960s, in a series of papers dedicated to sea urchin morphogenesis, Tryggve Gustafson of Stockholm University and Lewis Wolpert at University College London blamed the previous decades' "neglect" of research into cell movements and activities for the lack of progress in developmental physiology.[60] Proponents of time-lapse film, Gustafson and Wolpert asserted that cinematographic acceleration can cope with developmental processes so slow that they do not exist for us at all until they become visible "by virtue of their movement."[61] In classic papers little known today, they concur with Gräper that the great variability of cell movement is gastrulation's most salient feature. Though development is generally geared toward arriving at the same structures, nature has at her disposal a "variety of pathways" to reach that end. Gustafson and Wolpert conclude that the actual, specific "time-space distribution" of cellular activities must become the focus of research.[62]

As late as 1977, Michael Abercrombie (1912–1979), the pioneer of cell behavior and locomotion studies in postwar Britain, lamented that science had been "curiously slow" to accept the idea of active cell locomotion, which had met with "peculiarly strong resistance." Knowledge about the dynamics of processes and movements in morphogenesis had existed since the late nineteenth century, he pointed out, yet only rarely and incompletely had active cell locomotion actually been admitted as the explanation of an event. Despite the accumulated evidence, cell motion in embryology, wound healing, and cancer research had been overlooked due to a "mitotic infatuation" that insisted on interpreting cell proliferation as an effect of local mitosis.[63]

Time and Movement

In the historiography of science and film, it is the temporality shared by film and the living object that has dominated attention. Film and biology seem inextricably connected, to the extent that, in Hannah Landecker's words, time-lapse microcinematography facilitated a "distinct form of biological knowledge with time and movement at its center."[64] That common temporality is why film can capture the dynamism of life so convincingly and with such apparent ease. Gräper's research, for example, traversed all the stages of vitality, working with organisms that were dead, alive, temporarily prevented from dying, kept alive largely naturally, and experimentally manipulated with dyes and marks. If life is an event extended in time, then the cinematographic apparatus is a superb invitation to play with time: it can be condensed, stretched out, and speeded up and the film played forward or backward or stopped. Film undoubtedly and impressively made possible the "perception of life in time."[65] It put life's dynamics on display, introducing a visual counterpart to the scientific arsenal of static methods. Because of this "natural attraction" between film and scientific research, cinematography is widely credited with having paved the way for a new understanding of the biological world as, precisely, a living world.[66] With time at its core, film seems to bridge the "essential incompatibility or impasse between visibility and vitality," reintroducing something that had largely been lost and had made twentieth-century biology a "science of the non-living." Film reintroduced life.[67]

But even if Gräper's cells now could be made to live in cinematic time, they were not just moving. They were dancing. This book has followed life through motion, not through time. Movement is not so much that which reveals organic time in film — a basic technical prerequisite to make time apprehensible — as that which requires explanation. Film showed motion, but this did not help Gräper to grasp the role of motion in embryonic formation; only as a holistic pattern of movement that sustains itself under variable conditions did the cell movements he saw make sense. By playing with space and time simultaneously, he realized that the *choreography* of movements is what differentiates one fundamental event into many different gastrulation processes. It was seeing a dance that made him understand form.

Embryologists were able to investigate how motion creates form because microscopy, photography, and film could capture cells moving. As we have seen, however, it is not only cells themselves that move. Life is the activity of organic substance as a whole: the movement of cells, through cells, and in the interior of cells.

What happens to movement once it moves to the subcellular level? So far in this book, investigating biological motion has laid bare a view of life's agency as so vast, colorful, and rich that it can embrace the tipping point from not being to being, ensouled animals and frightening projections, the swimming and whirling of animalcula, and a graceful social dance in a courtly three-quarter beat. Given the potency of movement, the interesting question to ask seems to be not how moving images reintroduced life to the sciences, but whether movement's enormous scientific visibility in the twenty-first century actually takes something away from our understanding of life. Put differently: What does life become by becoming visible — what does producing motion in moving images do to life?

CHAPTER EIGHT

The Root of All Existence

Motion, previously marginalized, has come to center stage in these first few decades of the twenty-first century. Motion is now nothing less than the lever that seems likely to force bioscience off its hinges.

First and foremost, that is because motion is now at the core of organic life. On the subcellular and molecular level, in the inner life of cells where cargoes are delivered, molecules transported, and organelles displaced, movement turns out to be the "root of all existence."[1] Now it is cells that are "*doing*" life, and motor proteins are the agents of the newfound molecular mobility.[2] Second, whereas analog images of motion were not capable of storming the bastions of experimental science, their digital descendants have now captured that fortress from inside. Computer animations, live-cell imaging, and fluorescent tracking using high-resolution microscopy or simulations show the organism in maximal motion, full of internal agility — but they are also the *conditio sine qua non* for a world of research that can no longer think, calculate, or experiment without moving images. Third, motion is spearheading a robust critique of the established foundations of the biosciences. Most broadly, this critique proposes to conceive of life not as structure, but as movement, not as being, but as constant becoming. It originates, on the one hand, in theoretical biology and the philosophy of science as a processual philosophy that turns against the "metaphysics of things" so dominant in biological research.[3] On the other, it ensues from the related criticism of biology's mechanistic bedrock: the equation of organisms with machines.[4] Going beyond this critique, finally, are the efforts that have been gaining ground in different forms since the 1990s, initially in posthumanist and feminist Science and Technology Studies, to wrest the contemplation of biological life from the exclusive embrace of the natural sciences.[5]

In view of an unprecedented destruction of nature in the twenty-first century and the no less unprecedented possibilities presented by the scientific design of new nature, proponents of a "new vitalism" are raising their voices.[6] They counter fragmentation and analytical dismemberment upon smaller and smaller scientific territories with the broad scope of "togetherness."[7] This approach to the natural sciences, especially biotechnology, is fueled by a fundamental ethical unease. Having switched into the mode of making, thus intervening massively in the world outside the walls of academe, and having contributed through its methods and procedures to the twenty-first century's sense of crisis, science is increasingly under interrogation; doubt is cast on its role as the sole compass for the future order of things. From this perspective, nature as a scientific object comes to form the borderline between the sciences and the humanities. That frontier has to be negotiated afresh for every research question, since neither does science exist in "epistemological isolation" nor do its objects exist in "ontological isolation."[8]

If we assume that the living world is an event, new at every instant and embedded in an intricate braid of movements, changing relationships, and potential futures, then not only is it in constant flux, but no disciplinary approach, whether in the natural sciences, the humanities, or the social sciences, can continue to stake an exclusive claim to explain it. Nature meanders through the rich abundance of possibilities, disregarding human attempts at containment. To register life in motion means continually displacing the carefully tended palisades that corral the knowable.

In short, exploring life's specificity — the fact that it is mobile through and through — has prompted very different schools of thought to reconsider the limits placed on the knowledge of life. Put perhaps rather strongly, it has led them into the impasse of denying that knowledge can be specific at all. Both within and beyond purely scientific boundaries, motion has thus drawn twenty-first-century biology onto new and imponderable terrain. Yet strangely, the very opposite seems to be just as true: contemporary biology's intense interest in molecular mobility has transported it right back into the middle of the seventeenth century. Life now seems to be nothing other than the Cartesian universe of matter in motion. How did we get to this point?

Dynamic Instability

Until the late twentieth century, biologists knew that cells moved, and organic matter — whether it was called sarcode, protoplasm, or cytoplasm — was active and contractile. Under the light microscope, little more than that could be discovered about this fine, granular substance. What became visible under magnification were just a few structures, such as the nucleus, the cell membrane, or internal vesicles and vacuoles.[9] Visually, this mucus, gelatin, or albumen was condemned to insignificance as a "porridgy mass."[10] Its dynamic qualities, in contrast, began to interest chemistry at the same time as biology was turning its attention to the formlessness of organic matter. In the late 1830s, the Swedish chemist Jöns Jakob Berzelius (1779–1848) and the Dutch physician Gerardus Johannes Mulder (1802–1880) coined the term "protein" (from the Greek meaning fundamental or primary) to describe the macromolecules of which organic matter consists. Proteins, constructed of amino acid chains, could now be broken down into components, their structures and properties analyzed using the terminology of chemical reactions.

The "chemical behavior" of all organic matter, remarked the physiologist Wilhelm Kühne (1837–1900), was "unmistakably similar," a uniformity that applied most of all to the "contents of the muscle fibers." In chemical terms, the contractility of organic matter seemed to be comparable with the contraction of the muscle. Accordingly, it is with the muscles that Kühne, director of the chemistry lab in the pathology department of Berlin's Charité hospital under Rudolf Virchow, opens his *Untersuchungen über das Protoplasma und die Contractilität* (Studies on protoplasm and contractility) of 1864. After identifying the protein myosin as the central component of muscle, he turns to the motion phenomena found in all sorts of other tissues, whether plants, slime molds and amoebae, or the human cornea.[11]

The exact nature of chemical events in contractility, however, remained unresolved for almost another century. In the midst of the Second World War in Hungary, Albert Szent-Györgyi succeeded in distinguishing two different forms of myosin in muscle specimens. Szent-Györgyi and his colleague Bruno Straub also detected a further protein, which they called actin, and discovered that the muscle contracts when myosin and actin interact with the molecule ATP (adenosine triphosphate), the relevance of which to muscle contraction had become increasingly evident since the 1930s.[12]

The details of this interaction emerged under the coverslip of a microscope slide. In the mid-1950s, Hugh Huxley and Jean Hanson examined a suspension of myofibrils — long filaments, arranged in parallel columns, that form the muscle fibers — under the optical microscope. Photographing individual myofibrils on the slide while manually moving the coverslip and thus the myofibrils, they developed their sliding filament hypothesis. This proposed that during muscle contraction, actin and myosin filaments slide past each other, resulting in actin-myosin linkages and the contraction of the muscle. Seeking a "possible driving force," Huxley and Hanson suspected that ATP is split by the myosin, releasing energy.[13]

Actin and myosin were thought to be components of muscle cells alone until actin was found outside muscles in 1969, myosin in 1973.[14] Knowledge of other proteins in the cytoplasm was growing,

as well. In the 1960s, experiments on mitosis using the drug colchi-
cine uncovered microtubules—hollow, cylindrical protein structures larger than the actin filaments in the mitotic spindle. With
the help of chemical fixation using glutaraldehyde, it became clear
that microtubules and the protein, tubulin, from which they are
built are present throughout the cytoplasm.[15] In the 1970s and 1980s,
electron microscopy, rotary replication with platinum, and indirect
immunofluorescence made it possible to create visual representations of cytoplasm. The images showed rapidly frozen, "freeze-
dried cytoskeletons."[16]

The network of actin, myosin, and microtubules active in the
cytoplasm, now named "cytoskeleton," was the result of experimental, biochemical, and visual procedures: extracting the proteins, washing them out of their own environment, blocking them
with drugs, and making them visible in the electron beam. But
whether John E. Heuser and Marc Kirschner, writing in 1980, were
right to consider the cytoskeleton an "excellent representation of
the filaments in intact cells"[17]—in other words, how the representation actually related to the living material—could not be decided
unless that material became visible *in vivo* as well.

The late 1970s saw initial steps in that direction. D. Lansing Taylor and Yu Li Wang injected purified cellular actin, labeled with fluorescent probes *in vitro*, into living amoebae and tracked its chemical
reactions directly in the organism.[18] A new frontier of visibility was
reached in the early 1980s with video-enhanced contrast microscopy, in which a microscope paired with a video camera made the
movements of subcellular organelles visible *in vivo*. The technique,
devised by Robert and Nina Allen and their colleagues, produced a
greatly improved optical image of processes in the cell by viewing
them with a video camera that eliminated stray light. The price of
that technological vision was the loss of its human counterpart, for
the proteins now became "virtually invisible to the eye."[19]

Using this method, Allen was able for the first time to chart the
microtubules as they pushed their way into the field of vision on

his camera's monitor in bundles, separated out, and reconverged. Microtubules were filaments that stabilized the cell, yet they were also in motion. Experiments and visualizations *in vitro* additionally showed that they could increase or decrease in length and number, maintaining the cytoskeleton in a state of "dynamic instability."[20] The character of that dynamism, the inherent activity of organic matter, was still obscure. The world of molecules continued to elude the gaze and grip of science, but it now seemed certain to be bursting with movement.

Finding Movement

Cytoplasmic streaming, mitosis, intracellular transport — more and more processes were now identified as motion phenomena, but what was meant by motion here? Should it be taken to mean that contraction, flow, or transportation described different processes but the same movement? And who or what exactly moved? How, and to what end, should movements be distinguished or compared?

What molecular movement could come to mean biologically depended on the experimental systems by means of which it was produced. Experimental systems, as Hans-Jörg Rheinberger has observed, are "complex, tinkered, and hybrid settings" for the "emergence, change, and obsolescence" of scientific objects. In the routines of experimental manipulations, scientific objects are not stable; they constantly appear and disappear, subject to permanent alteration or complete reconstitution by the scientific procedures, visual representations, and surprises that the experimental setup is designed to generate.[21] This "generation of differences" is the "reproductive driving force" of the experimental system and of its mission to bring about "scientific novelties."[22]

For molecular movement to become a novel scientific object, organic matter had to be both alive and visible — two parameters that are articulated in different ways in the experimental setups of the twenty-first century. A hundred years ago, in Ludwig Gräper's studies, technologies to make living substance visible and keep it alive enabled science to be done differently. In the hybrid settings of today's biotechnology, however, being alive and being visible are no longer poles on a spectrum within which only limited adjustment is possible. The organism is no longer preexisting or produced, alive or kept alive, real or represented; rather, it is inseparably both at once.

Despite or because of these achievements, molecular movement posed challenges to science that went beyond the new technologies of motion generation. To make molecular movement a scientific object, imagination was required. In a rather old-fashioned way, making molecules move required scientists to fit nature to their imagination. And that imagination, not surprisingly, was cast in a very long tradition.

Movement in a Test Tube

According to the sliding filament hypothesis, myosin moves over actin in muscle contraction. These actin-myosin movements, previously only conjectured, were first made visible in an experiment with the alga *Nitella*. Michael Sheetz and James Spudich washed the alga's cytoplasm out, exposed the actin filaments, and applied myosin-coated fluorescent beads. Watching the beads proceed along the actin filaments, they found a "striking directed movement."[23]

In 1985, Sheetz and Spudich standardized their procedure and designed the first experimental system to investigate motion, an *in vitro* motility assay. Myosin-coated fluorescent beads were anchored to a scaffold of purified muscle actin adhering to a glass slide. With optical microscopes and time-lapse photography, the assay showed the fluorescent beads (and thus the myosin) advancing in a linear direction along the actin filaments.[24] In the controlled setup of the assay, motion made visible becomes a quantitative instrument. For example, it enables the motion's velocity to be measured and conclusions to be drawn from those measurements regarding energy consumption or the molecular structure of the proteins.

In the same period, video-enhanced contrast microscopy revealed busy activity and organelle transport in the axons of nerve cells. Organelles displace toward or away from the cell body through the axon.[25] Having worked on *Nitella* and the motility assay, Sheetz and Ron Vale, a doctoral student at the time, revisited these movements from a new perspective: they equated the motion of the organelles through squid axons, observed *in vivo*, with the movement of the beads in their motility assay, produced *in vitro*. From the optical similarity of transportation and muscular motion, they inferred the identity of the biochemical processes, surmising that actin and myosin are also involved in organelle transport inside the nerves.[26]

The starting point for further experimentation in this case is the observation of a directional movement of particles in the cytoplasm, whether *in vivo* or in an assay.[27] Despite being used only as protein markers, the beads in the assay are now treated as goods in transit, equivalent to the organelles. This step is what opens the way to the hypothesis that one and the same protein, myosin, must be involved in both processes and to the conclusion that in the filaments that dictate the direction of movement, again one and the same protein must be at work — in this case, actin. An inference is thus drawn from visualized motion to the underlying movers, the proteins, and in order to draw that inference, the elements of the assay and the cell are regarded as interchangeable. This dissolves the dividing line between organism and assay while also making it possible to treat contractility and organelle transport as identical processes.

In such experimentation, the underlying assumptions as regards motion are, first, that structures exist by means of which motion can be made visible; they are the agents of motion. Second, because these movements are not random, there must be structures to which the motion is bound — that direct it as "carriers" of movement and regulate it when necessary, for example, when organelles "fall off" their tracks or remain motionless.[28]

The researchers had assumed that these carriers were actin. Through the electron microscope, however, the filaments proved to be microtubules.[29] This meant that the proteins involved could be neither actin nor myosin, since these always appeared in combination, and there was no known protein that associated with microtubules. In addition, the beads in the assay were only visible markers of movement — they did not perform movements themselves.

To continue along the path of the assay in the face of these results, its logic had to be reversed. Initially, the living system of the squid axons had been equated with the contractility assay, permitting an extrapolation from nerve cells to muscle cells and accordingly from the motion of transport to that of contractility. This equation was now turned upside down. In the assay, the beads did not move of their own accord; their movement was caused by the myosin enveloping them. Accordingly, the organelles must likewise not be the cause of their own movement, but also be moved by proteins. Although the researchers knew of no protein that did any such thing, one must exist, and the assay would bring it to light. Whereas in the first case, the assay's design had worked to engineer movement out of known elements, the proteins actin and myosin, now the movement generated worked to demonstrate the existence of its elements, one of which was still unknown.[30]

Sheetz and Vale applied cytoplasm containing the as yet unknown protein onto a glass slide. In their new assay, delicate, dark lines could be seen gliding over the glass surface. This movement would give the assay its name, the microtubule gliding assay. For comparison: in the first motility experiment, the bead assay, individual, isolated beads could be seen moving in a straight line, as if threaded on a string of pearls, along a substrate that was not further specified. Their movement was "directed," "rapid," without "tumbling" though with "changes in velocity."[31]

Both assays reveal forms of motion — the motion of round, isolated elements and that of longer, rodlike forms moving over a rather undifferentiated substrate — but each interprets motion in a completely different way. In the myosin case, motion becomes visible in the manner in which the protein itself moves; the gliding assay upends that, showing how the microtubules, as transport filaments, are moved by the unknown protein.[32] This is why in the first assay, conclusions about the protein could be drawn directly from the movement, whereas in the second, the apparent movement of the microtubules indicates only the concentration of the protein. If the concentration is high, the microtubules' motion is restrained by the number of proteins attached to them and proceeds without serious perturbations; if the concentration is low, the motion is erratic.[33] The motor protein detected in the cytoplasm of nerve cells in 1985 by this means was given the name kinesin.

The kinds of assays I have described are still standard in the investigation of molecular motion.[34] They shape motion according to a basic template that goes all the way back to Aristotle: motion is pinned down in one part so as to make it visible, and thus apprehensible, in another part. Assembled *in vitro* from their constituents, proteins are made to react biochemically under controlled conditions in order to produce motion. Kinesin becomes the "translocator" or "molecular motor" in the "transportation" of cell organelles,[35] and myosin, the muscle protein, can trigger contractions by "walking along" actin filaments.[36] Motion, here, is distributed between different actors — the proteins that move and those that are moved, or along the length of which movement takes place. It now has its first images, components, and structures. But why can proteins walk, fall, rest, and tumble at all?

Walking Kinesin

Proteins are macromolecules that consist of amino acid chains. They are active organic forms — or, rather: they are forms in action. Their biological potential unfolds through their distinctive, endless combinatorics of composition, spatial structure, and activity. Chemically, the proteins actin, kinesin, and myosin are enzymes, ATPases that, as biological catalysts, are involved in the process of splitting ATP into ADP. This chemical reaction, dephosphorylation, releases energy. Actin, myosin, and kinesin are therefore known as motor proteins or engines, since in the macroscopic world, the conversion of chemical energy into mechanical motion is the function of a motor. Motors are defined as machines with the goal of performing mechanical work.

In chemical reactions, the proteins' structure and their dynamics change simultaneously, so that if scientists wish to decipher organic processes, they must weave together several levels of analysis at once. The technologically realized observation of protein movements, the quantitative analysis of chemical reactions occurring, the attribution of proteins' structural states to particular phases of the chemical reaction or of activity, the attribution of particular movements to the functional context of the cell and its vital tasks (such as metabolism, cell division, or immune response), the description of mechanical activity as concrete movements—only when all these elements are correlated under the controlled conditions of the assay can the protein become a scientific object. Then, kinesin can walk.

The assay that identified kinesin showed how kinesin moves the microtubules, dependent on its concentration. Subsequent experiments with different concentrations of kinesin molecules showed that a single such molecule is capable of moving a single microtubule.[37] This made kinesin especially well suited for the further quantitative and visual investigation of motion on the level of single molecules.

Chemically, the hydrolysis of ATP to ADP releases energy; structurally, it results in a shape change—a conformational change—in the protein. By calculating the speed of microtubule movement captured on video while controlling the number of kinesin molecules in the solution, researchers found that kinesin is not continuously bound to the microtubule. Instead, particular cycles of ATP hydrolysis correspond to periods of kinesin attachment to the microtubule, whereas in other phases of the reaction, they are separated. In the controlled setup of the assay, it is possible to determine quite precisely the periods of attachment and the distance that the kinesin covers before reassuming contact with the microtubule.[38]

From a chemical point of view, then, kinesin and microtubules are periodically bound or not bound to one another; from a mechanical point of view, they execute a discontinuous local displacement, "dwelling for times at well-defined positions on the substrate, interspersed with periods of advancement."[39] The conformational change that occurs when phosphate is released is also described in terms of a "power stroke."[40] A few years after the assay experiments, investigators succeeded in tracking kinesin movement directly at molecular resolution. The power stroke acquired a visual equivalent: "kinesin stepping."

Working with a laser interferometer and optical tweezers, it is now possible to capture a single kinesin molecule optically. Displacements in space and time can be measured precisely, down to the nanometer and millisecond. This is done using a mathematical method, the pairwise distance distribution function, which calculates the likelihood of finding one particle of a pair within a certain distance of the other in a given volume. For the discontinuous motion of kinesin, the calculations gave rise to an engineering-based model: kinesin moves in discrete steps.[41]

In this model, the protein's conformational change is now one step, and that step is the distance traveled between dwell states as the kinesin makes "one step per hydrolysis (or perhaps fewer, requiring multiple hydrolyses per step)."[42] Active kinesin stepping conforms chemically with the molecular structure of the microtubules. These consist of tubulin, a heterodimer composed of negatively charged ⊠-tubulin and positive ⊠-tubulin proteins. To form the walls of the microtubules, typically thirteen such dimers assemble in a ring as a protofilament. Because tubulin units are added and lost at their ends, the microtubules are dynamically unstable, constantly growing and decaying. In the cell, they are usually bound at their minus end to a centrosome, the cell's microtubule-organizing center (MTOC), from where their plus end grows out toward the cell membrane. The kinesin moves in parallel to the protofilament axis and binds only to the ⊠-tubulin, so that the kinesin's stepping distance will always be a multiple of the distance between individual tubulin binding sites on the microtubules.[43]

This engineering model, reliant on the exact correspondence of biochemical and mechanical processes, was by no means uncontroversial in the 1990s. One alternative hypothesis was that the kinesin is folded in its form. And whereas the assays I have described disregarded the role of the cargo transported, other studies suggested it might be precisely this cargo that activates the kinesin's movement, so that "caution should be exercised in mechanistic interpretations of kinesin action in both *in vivo* and *in vitro* motility assays."[44]

If we do, though, assume that motor proteins are machines, albeit "very unusual machines that do what no manmade machines do,"[45] how exactly does ATP hydrolysis interlock with movement, and specifically with stepping?

The answer is to be found in kinesin's structure: it moves on two heads. Unlike tubulin, kinesin is a homodimer, composed of two identical proteins. Electron microscopy and chemistry identified a structure that has become known as kinesin's head, stalk, and tail—or more precisely, a group of seven different functional domains, the most important being a tail that interacts with other proteins or cargo, a long, helical coiled-coil domain in the center, and a flexible neck connecting this to two globular head groups at one end.[46] These groups are kinesin's "motor," its "head," its "feet."

As it moves along the microtubule, a kinesin molecule's motors cyclically bind and unbind, but the molecule remains connected to the track for many seconds or hundreds of steps in its processive movement. The observation that the molecule is simultaneously both attached and detached permits essentially two hypotheses as to how the kinesin heads move: either trudging "hand-over-hand" or creeping along like an inchworm. In the first case, the two heads swap their roles at every step; in the second, it is always the same head that shifts forward while the other follows it. In both cases, one head always remains attached while the other moves forward, and the movement arises from the coordination between the two heads.

Each head has a binding site for the microtubule and another for the ATP, and the homodimeric structure of kinesin suggests that the structure and movement of the heads may be equivalent, consequently that both "have the same hydrolysis cycles and make the same motions."[47] This is far from clarifying the gait of kinesin, since step equivalence does not mean head equivalence. If the steps of the two heads are the same in every respect — in other words, if the heads take turns in movement, chemical hydrolysis, and attachment (the symmetric hand-over-hand model) — then the moving head must always swing forward on the same side of the attached head. But that would imply that the stalk, the coiled-coil domain of the molecule, twists more and more at every step, which is not the case. In the inchworm model, the steps are also equivalent, and the kinesin also covers the same distance with every step, but the two heads do not behave identically: the leading head moves from one binding site to the next and waits there for the trailing head to catch up. One head is catalytically inactive. Yet another possibility is that the heads are equivalent, but their steps are not. The kinesin stalk does not twist because the lagging head overtakes the leading one on a different side each time, so that the torsion is negligible and can be compensated by the flexible neck linker domain — the asymmetric hand-over-hand model.[48]

It is this third, asymmetric model of kinesin stepping that makes kinesin "walk." The model does not entirely resolve the issue of how chemical and mechanical steps are coupled to enable coordinated movement. Indeed, the asymmetry of the hand-over-hand model raises the question of whether kinesin does not actually walk, but limps. The "limping factor" describes the problem that one head moves faster than the other, due to the asymmetry between odd and even steps and the structural difference that the motor reverts to exactly the same state only after every second step.[49] Despite this weak point, the asymmetric hand-over-hand model made "so much intuitive sense that it has found entry into all major cell biology and biochemistry textbooks."[50]

Do Engines Walk?

So is kinesin "the world's tiniest biped," which "walks like a person?"[51] Specifically, like a person "walking across a pond along a row of stepping stones," heads thrusting forward "akin to a judo expert throwing an opponent with a rearward-to-forward swing of the arm"?[52] Does the protein really share with human beings "all the hallmarks of bipedal walking," or, with its two identical heads, does it not strictly speaking have "two left feet"?[53]

Fish swim, snails creep, birds fly — we distinguish movements in part by the physical conditions in which they occur. Proteins, though, all move in largely one and the same organic medium. More than that: their movement seems to resemble the human gait, which Ovid describes in the "Creation of the World," the first book of his Metamorphoses. There, new-made man, "while the mute Creation downward bend / Their Sight, and to their Earthy Mother tend," possesses a different privilege and responsibility — he "looks aloft; and with erected Eyes / Beholds his own hereditary Skies."[54]

When engines walk in that same way, it has a paradoxical charm. Machines arose in the seventeenth century as imitations of particular types of motion that human beings could not or did not want to perform themselves and preferred to delegate to an inanimate object. We encountered the peculiar logic of seventeenth-century mechanist thinking in the case of Descartes, discussed in Chapter 1. In Descartes's world, the machine was invented as the imitation of a living model; however, that imitation of a living model does not copy the living model, but itself comes to model the natural phenomenon. The machine, a material construct, takes on a clearly epistemic task: it counterposes the cognitive faculty of human beings with an organic nature of which the machine itself is not part. For Descartes, the machine manifests the *res cogitans*, a rationality that is not part of nature, yet is required in order to understand nature.

In the twenty-first century, the Cartesian dualism of body and mind has had its day. Mechanist thinking in biology, too, has long since broken away from those origins, yet the machine is still a dominant epistemic model, human reason's order for the structural and functional relationships of nature.

Our particular iteration of the machine, the motor, can convert energy into many kinds of motion, but walking and limping are not customarily among them. In fact, in the case of the motor proteins, this particular form of movement and the machine as its producer seem not merely incompatible, but mutually exclusive. For human reason, the notion of the motor here works as a way to order movement of a kind that — culturally, historically, and mechanically — is the antithesis of machine motion.[55] Added to that, the motor no longer confronts nature as a model: it is now physically incorporated into nature, a powerhouse in her invisible depths.

Talk of limping and walking proteins articulates, on the one hand, the desire to apprehend motion as the correlation of proteins' dynamics and structure; on the other, the still-unresolved problems and assumptions upon which the engineering model rests. If motors walk, then the machine has become part of a nature that it now *is* and can no longer *explain*. Descartes seems to have been turned entirely upside down: the machine is now part and parcel of nature — motor molecules *are* life — but machine rationality no longer fits nature. Neither does nature seem to require it any longer, for motor molecules do what no manmade machines do. In this sense, walking kinesins and limping motors point to the elephant in the room: the question of why organic matter can move of its own accord at all.

Animating Science

One way of tackling that elephant is through the image. Among both scientists and the wider public, the kinesin's walk quickly entered the limelight in book cover illustrations and millions of YouTube clicks. Its iconography was greatly influenced by an eight-minute computer-animated clip produced in 2007 for Harvard biology students, "The Inner Life of the Cell" (fig. 8.1).[56]

Figure 8.1. Walking kinesin, animation created by BioVisions at Harvard University and the scientific animation company XVIVO (2006).

Figure 8.2. Stills from "A Day in the Life of a Motor Protein," made at Utrecht University in the molecular neuroscience laboratory led by Casper Hoogenraad, 2013, www.youtube .com/watch?v=tMKIPDBRJ1E.

Among other things, this animation shows a wispy structure, delicate and pliable as a sapling, pacing sure-footedly like a high-wire artist walking a taut rope. Instead of carrying a balance pole, it tows behind it a gigantic sphere. Despite the weight of the orb tied to the dainty walker's branching ends, dragging it back, it still launches its oversized feet forward in springy steps to pull its burden onward. Other parts of the animation give glimpses into the space of the cell, which is dimensionless in its void, yet also full of dynamic abundance and crammed with life.

In 2013, the laboratory of the Dutch cell biologist Casper Hoogen-raad at the University of Utrecht animated a day in the life of the kinesin "John" (fig. 8.2).[57] John is a tall, slender fellow with one big eye, an expressive mouth, and giant slippers adorned with purple pompoms. Woken by the alarm clock to a day of duty, he mingles with real city life as he walks through the streets of Utrecht. No differently than the cyclists, pedestrians, and buses bustling around him, John, too, moves with a purpose: to deliver cargo. But just as for them, carrying out that task is not straightforward. Roads or lanes are blocked, streets are crowded, traffic is regulated by police, and antagonists are out to meddle in John's business. Dynein, his companion motor protein, insists on walking the opposite way, and because "there can be only one direction," John has to enter a tug-of-war.[58] When the traffic police clear John's freight to move ahead, he faces yet another obstacle: the road is under construction. He is not alone. There are many more Johns and much more cargo to be delivered through the streets, popping up at different crossroads and sometimes joining paths at different times and with different degrees of urgency. Ultimately, only John can ensure a smooth flow of traffic, enabling the city to function and the organism to thrive.

The Harvard cell animation portrays events in the cytoplasm with the colors and swaying motions of an otherworldly underwa-ter realm, whereas Utrecht John's movements are integrated into real-life footage of the city on a summer's day. Both animations have been immensely successful since they were first posted, but they and other animations have done more than contribute to the vivid presence of walking motors in the popular imagination. In recent years, they have also attracted considerable attention in science and technology studies and in media studies.[59]

Natasha Myers's ethnographic study on the work of protein crystallographers, for example, shows that they see molecular life not as a set of "fully deterministic machines," but as "rather untamed." Accordingly, they describe their work "in registers that waver between the machinic, the human, and the animal."[60] Animations belong to an important register of this kind, one that works with both effects and affects. Myers emphasizes the role of performativity and embodiment, which are accorded no place in mechanistic conceptions of the protein, yet are fundamental to their formulation.[61] Animations such as the Harvard cell film are "embodied animations," expressing scientists' fantasies and intuitions, a "*feeling for* protein forms and movements."[62] A modern form of enchantment, they are proof that the "animisms and vitalisms of the earlier sciences of life" have never been completely erased.[63]

Animators, in turn, have compared their work to that of biologists. They see themselves, as Alla Gadassik writes of the Russian-born French animator Alexei Alexeieff, as "microscopists of the cinema." Animations share epistemic ground with the life sciences, since both biologists and animators seek the core of their own science or art in motion. By constructing motion in their images, animations create an illusion of vitality not only at (or beyond) the margins of the living world, but also in the midst of the sciences, where they bring back to life a world that has so often been stilled by instruments, analyses, and modeling.[64]

In the postgenomic life sciences of the twenty-first century, such "life as still life" has become a thing of the past. Contemporary biology has a new vitality, thanks to imaging technologies that make it possible to observe processes as they take place in living cells by labeling cells and subcellular elements with fluorescent markers, capturing their behavior with high-resolution microscopes, and processing that data with computer technology.[65] Biology has become a science of structures in motion, an exponentially growing database of imagery that includes everything from molecular motors tagged with gold nanoparticles, to fluorescent cancer cells metastasizing into healthy tissue, to proteins navigating the city cell.

Unlike their forebears in scientific cinematography, the new scientific worlds of moving pictures, from the Harvard cell movie to live cell imaging, have a tremendously high profile, reaching a broad audience in the radically changed landscape of digital scientific publications and the technological infrastructure of the internet. Supplementary materials, *in vivo* visualizations, and live cell movies have become firmly established, openly available on platforms such as YouTube or taking pride of place on the websites of important journals.

The concept of animation points the discourse on contemporary biology toward the epistemic relevance of biology's visuality, and it does so by means of a bedazzling mediation between motion and aliveness, construction and performance, structure and relationships. Biology, argues Hannah Landecker, constantly oscillates between what it can show and what it can grasp theoretically. To do justice to its object, it relies on capturing biological vitality in images and is thereby locked into a cycle where "theory animates observation, and then observation is preserved, quantified, pinned down to make the perceptible intelligible within scientific conventions of (objective) proof, evidence, and demonstration. But then each round of codification is met with some dissatisfaction for its lifeless representation of life; a machine is built to animate observation's codification, and the resulting moving image is perceived as an animation of theory."[66]

That theory, forging ahead on the computer screen, is life as a molecular network. Just like the gene in the twentieth century, in the twenty-first century it is the molecule in interaction that drives biological research forward. The theory of life as a molecular network is interpreted by some as a new form of vitalism — a molecular vitalism.[67] But vitalism or not, making images move is a way of performing knowledge, and the question of how that performativity affects epistemes is a promising one. Primarily, though, moving images is a way of performing motion. Beyond direct observation, motion is elusive; it exists only in reconstruction.

It was the special way of generating motion that the Canadian animation pioneer Norman McLaren, working in the 1960s, saw as the art of animation: "animation is not the art of drawings-that-move, but the art of movements-that-are-drawn." The representation *on* each frame is thus less important than "what happens *between* each frame."[68] Film scholar Tom Gunning identifies that epistemic switch as the fundamental difference between film and animation. Film, he argues, presents "movement automatically captured through continuous-motion picture photography," whereas animations consist of "images that have been artificially made to move."[69]

Life is movement, but a science of life rests on knowledge of how we make that movement. The concrete manner by which movement is generated is the key to understanding it. After all, what is fascinating about animation is its ability to construct motion where there is none. If projecting sunlight made protozoans into monsters and microscopy turned cells into dancers, what will nanoscopy do to the roots of all existence?

Single-Molecule Visualization

Visualizing movement on the nanoscale is crucial to comprehending how enzymes operate. Not surprisingly, imaging technologies have supplied excellent candidates for a Nobel Prize since the turn of the millennium. In 2017, 2014, and 2008 the Nobel Prizes in Chemistry were awarded for cryoelectron microscopy, superresolved fluorescence microscopy, and the discovery of green fluorescent protein (GFP); the 2003 prize in medicine or physiology honored the development of magnetic resonance imaging; and in 2018, the physics prize went to the invention of optical tweezers and their application to biology.

When Eric Betzig, Stefan Hell, and William Moerner, recipients of the 2014 Nobel Prize in Chemistry, used superresolution microscopy to transcend the limits of optical microscopy, they brought visualization into the realm of nanoscale resolution. At the same time, high-technology imaging procedures not only prompted Betzig's much-cited question "When can we believe what we see?" but also caused viewers to wonder exactly what "seeing" should be taken to mean in this context.

Initially, motility assays such as the bead and the microtubule gliding assay brought cytoskeletal motor proteins into the scope of the petri dish. As early as 1993, Karel Svoboda and his colleagues announced a first "direct observation" of kinesin's stepping in the title of their *Nature* paper. To be sure, what they were able to observe was not the protein itself, but beads attached to it. The scientists combined a dual beam interferometer with optical tweezers. Optical tweezers or traps work with optical forces, but they are not an imaging device — rather, they act as a kind of linear spring. Prisms split laser light in a microscope into two beams that function as pincers. In the assay, silica beads with kinesin molecules bonded to their surface are caught in the focus of the trap. The kinesin molecules escape from the focus when they move, detaching themselves from the microtubules that are fixed on the assay coverslip, but they are pulled back into the trap again and again. During these "multiple cycles of attachment, movement and release," the forces that arise are measured.[70] By detecting the various kinesin positions in the trap and analyzing the data mathematically, Svoboda's team drew inferences regarding the type of kinesin motion, which they found to be a discontinuous stepping. "Direct observation" of kinesin stepping, therefore, does not mean that either the kinesin molecules or their steps can be seen directly. Instead, this is a statistical analysis of data, generated using optical means, about the location of the silica beads.

In order to target the kinesin heads—which are many times smaller than the beads—directly by optical means, single-molecule detection techniques are required, using microparticles and nanoparticles such as gold and quantum dots or fluorescents as optical probes. This burgeoning field is constantly giving rise to new methods, including combinations of single-molecule, high-resolution, and superresolution microscopy with acronyms recalling "Hollywood animation characters"—single-molecule high-resolution colocalization (SHREC), fluorescence imaging with one-nanometer accuracy (FIONA), defocused orientation and position imaging (DOPI)—along with laser dark-field illumination, adaptive optics lattice light-sheet microscopy (AO-LLSM), and single-molecule fluorescence (smFRET).[71]

FIONA was developed to resolve the question of kinesin gait for the cases of myosin and kinesin.[72] A single kinesin head is tagged with a fluorescent dye and tracked with a special optical microscope, the TIRFM (total internal reflection fluorescence microscope), in order to observe whether the heads move hand over hand or like an inchworm. TIRF microscopy is one of the most frequently used single-molecule detection procedures. Incoming light at a shallow angle is totally reflected at the interface between the coverslip and the aqueous sample; microscopic observation is restricted to this thin interface and the optical interferences that occur there. The method minimizes background fluorescence and confines fluorescence emission to the interface layer, enabling a higher optical resolution.

The intensity of the fluorescence emitted by the motor protein is detected using high-resolution cameras. For the further procedure, it is essential to localize the position of the molecule precisely, which means determining the center of the fluorophore in the particle image. In optical systems such as TIRFM, this is done by mathematical means: the point-spread function to describe the intensity distribution of an image and a Gaussian function calculated from various parameters, including the number of photons collected, the number of pixels and pixel size of the detector, and the standard deviation of the background. In this way, a resolution higher than the diffraction limit is achieved, making it possible to distinguish particles such as the two kinesin heads at the nanometer scale.

To follow the position of the particle over time and space, tracking algorithms are needed. Commercially available software (such as ImageJ or FIESTA) is generally used, first to convert grayscale into binary images in order to identify objects above a certain intensity threshold as particles and to localize them individually in each image. The particles' trajectories emerge when particle positions from individual frames are then tracked and linked across consecutive frames.[73]

The temporal resolution of the movement in this experimental system depends on the rate at which the images are captured. In 2016, using a similar microscopic setup, but gold particles instead of fluorophores, Hiroshi Isojima and his colleagues were able to monitor the position and time traces of kinesin at a higher temporal resolution and thus detect not only kinesin's steps, but also their intermediate states.[74]

The high spatiotemporal resolution achieved in this research tempted the scientists to present kinesin's gait in an iconic pictorial guise. Their portrait had kinesin walking through a series of images with a grid backdrop, one of the hallmarks of Eadweard Muybridge's work (Chapter 2). Inspired by Muybridge's famous series *Sallie Gardner at a Gallop*, Erwin Peterman in *Nature Chemical Biology* presented a cartoon strip in which the first and last image show the step's start point and end point (measured at 0 and 16 nm), the four images in between showing the intermediate states (fig. 8.3). Blurring the appearance of the paler stepping head (red in the original) adds to the impression of active motion.

Here, the historical reference equips kinesin stepping with a visual self-evidence that appears to *demonstrate* single-molecule movement, just as Muybridge's chronophotographs once demonstrated the horse's gallop — even though Isojima's experiment was

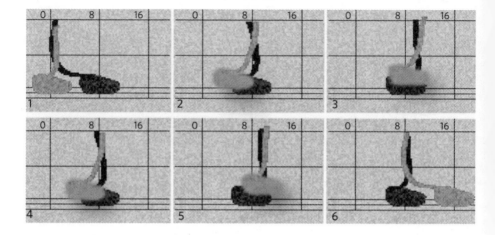

Figure 8.3. Cartoon depiction of the gait of conventional kinesin. Blurring of the moving red (paler) head indicates the intermediate states. Peterman, "Kinesin's Gait Captured," 2016.

designed to gather data and these data came from a single, labeled kinesin head. Peterman's fascination with the experiment arose not least from the fact that "the data shows it all, without complex data analysis." Images, apparently, do not count as analysis.[75]

However, trapping and tracking single molecules in an assay is one thing, observing them *in vivo* is another. *In vivo*, external manipulation of the cell is minimal, but natural disturbances are maximal. The cell is crowded, full of autofluorescence generated by the cell organelles and brimming with movement in all directions, at all speeds, and by all kinds of actors. Tracks cannot be passivated and flattened onto a surface, neither can motors be easily singled out or slowed down or the physiological environment decomposed and controlled. For all these reasons, more advanced single-molecule techniques *in vivo* have become highly sought after in recent years.

Naturally, the cell is inhabited by an enormous number of molecules. One way to increase the visibility of the specific molecules of interest in such an environment and to access them for the purposes of measurement is to refine the performance of the reporter particles. Among the various techniques to achieve this are quantum dots. These semiconductor nanocrystals are much brighter and more photostable than organic dyes, though they have to be applied to the molecule *in vitro* before being transported into the cell.[76] Using photoswitchable fluorophores as labels, microscopic technologies such as STORM (stochastic optical reconstruction microscopy) or FPALM (fluorescence photoactivation localization microscopy) activate one subset of fluorophores at a time in order to track the trajectories of larger numbers of molecules in real time.[77] Single-molecule fluorescence resonance energy transfer (smFRET), too, permits the measurement of nanoscale motion by exploiting the energy transfer from a donor to an acceptor fluorophore. In molecular motors, for example, the fluorescent labeling takes place in two different domains of the same molecule. If the pair of labels gets close enough, energy is transferred and the acceptor molecule emits a photon; this can be detected and tracked to measure the distance between the parts of the molecule.[78]

As optical technologies and registering devices become more inventive, sensitive, and closely tailored to particular samples and research questions, technology aspires to outwit the messy physiology of the living by laying optical siege to invisible subcellular life. Ultimately, the scientists' aim is to combine tracking and monitoring methods, ensemble and individual measurements, to obtain a portrait of the cell's inner life in glorious technicolor. Labeling simultaneously with different colors and distinguishable fluorophores can bring to light not only cytoskeletal tracks, molecular motors, and those molecules' cargoes and binding partners, but also the ways in which cargoes and organelles are transported, motors interact in the dense environment of the cell, and structures take shape in space and dissolve over time.

"Lighting up life," to quote Martin Chalfie, the 2008 Nobel Prize-winning discoverer of the fluorescent protein GFP, has effectively become a game of optical tactics.[79] It is an artifice to divide and conquer natural complexity by switching actors on and off at will, illuminating some and assigning others to obscurity; choosing, varying, and layering levels of complexity and interaction — like sending John the kinesin off to walk the streets of Utrecht or setting the nimble kinesin walker on its track, or, to continue the metaphor, like placing isolated, lurid signposts of knowledge amid the vast expanse of darkness.

Biology's Next Microscope

Leeuwenhoek watched the movements of animalcula through his bead lenses, but his depictions with pen and pencil failed to make his readers participate in seeing what he himself observed. Gräper pointed his camera lens at cells, but when registering their movements, he was dissatisfied, missing the resolution in depth that would have been required to convey his perception of a dance. Today, the new microscopy musters impressive experimental skill and abundant optical and biochemical knowledge in order to draw the nanoscale into the visible world.

Because it enables us to perceive the hidden level of life as a colorful event in cinematic time, computer animation is considered an imaging revolution in contemporary biology. That imaging works in a very particular way: the molecular movement we are made to see is neither stilled nor recorded, but calculated. Computer animations thus certainly constitute a novel addition to the technological perception of motion, but they are perhaps more revolutionary in their impact on the dynamics between biology and mathematics. In the words of Joel Cohen, a Columbia University mathematician and population biologist, twenty-first-century biology and mathematics have entered a new phase as "an explosive synergy" that is "poised to enrich and extend both fields."[80]

As we have seen, in high-resolution microscopy, mathematical modeling turns the emission of the fluorescent labels attached to the molecules into pixel data that algorithms then identify as objects, the trajectories of which they compute through consecutive image frames in time and, finally, fit into diagrams, histograms, or animations. To model movement mathematically, differential equations are needed. Ever since Newton and Leibniz developed the theory of infinitesimal calculus at the end of the seventeenth century (Chapter 1), calculus has been the mathematical means of describing motion and change. A differential equation relates to a set of functions with one or several variables; the equation describes how the variables change in relation to one another. When differential equations are exactly solvable, their solutions can be expressed in terms of integrals. More often, differential equations are partial or nonlinear, in which case they have no exact solutions; solutions to a given differential equation can only be approximated numerically, with the help of the computer.

In principle, differential equations can be applied to many events in the physical world, but biology poses particular challenges. These arise partly from the heterogeneity and complexity of the problems encountered in organisms and partly from the fact that algorithmic solution procedures for biological problems such as molecular movement are more advanced in computing practice than in their theoretical, mathematical foundations. From the mathematician's point of view, however, it is precisely these intricate challenges that promise to "stimulate the creation of qualitatively new realms of mathematics."[81] And from the biologist's point of view, the fact that mathematics is needed in order to interpret optical data makes it "biology's next microscope." If anything, it surpasses the microscope because of the universal application of mathematics to the interpretation not just of optical data, but of any kind of data at all.[82]

To make the invisible move under the conditions of visibility, in short, it takes experimentation, mathematics and computation. Animation, in this sense, is mathematics made to move. But when animation and mathematics give form to data, what does the study of movement gain? As we will see, they bring into focus what has hitherto been strikingly absent: the possibility of comparing movements.

Movement Rules

Comparing *images* is part of our cultural literacy. We know how to look at them so as to discern differences, single out elements, or identify the figurative means chosen to depict objects and create spaces. We are skilled in locating them within a long history of other pictorial references and can see how images move and morph as time passes and contexts change. Ludwig Gräper, too, began his motion research by comparing images and measuring distances with a ruler in order to trace the displacement of cells.

How, though, do we compare *movements*? We are fully aware of the difference between a crawling snail and a horse in full gallop, but what distinguishes two snails crawling or two horses in gallop — let alone two molecules in motion? As a rule, watching movements does not suffice to tell us whether they are exactly the same, almost exactly the same, or only rather similar. Furthermore, the formalization of a process in terms of signs or mathematical descriptions is a representation of a process or a set of instructions for a process; it is not itself a process. It requires some kind of actualization or instantiation.

This raises two questions: How can we make movements comparable with each other, and upon what basis can we compare processes generated in a representation with processes taking place in nature? Above and beyond those questions, Gräper's work indicates a different scale of comparability. Gräper understood that even if a common goal exists, the ways of moving toward it are manifold. Motion is a trajectory in its own right, so it is not possible to infer from the product of a process the motion that brought it about. Additionally, the extraordinary diversity of early developmental events in different animal species, Gräper showed, does not mean that they are not all the same in principle. Rather than being distinct forms of gastrulation, they are based on just one choreography. By shifting attention to the wholeness of movement, Gräper was able to apprehend the commonality of divergent gastrulatory movements across species — in the pattern of a dance.

The dance's pattern escapes cinematography, because although movement may be displayed in the image, it is comprehended through the rules that it obeys. This was the distinction that helped Johansson to revolutionize the notion of perception in the 1950s. Perception, he claimed, is not "factual," in the sense of having its equivalent in a pictorial representation, but "plastic," guided by potentially multiple forms of mathematical description.[83]

Mathematics is, of course, the most coherent system of formalizing rules. Describing movement in terms of mathematics gives continuing change an internal coherence and comparability as a rule-based occurrence. But to describe the rules obeyed by movement mathematically, no less than in any other way, one has to choose a level of contextualization, of relationality, regarding matters such as the internal coherence of the movement or the scale of comparability between movements. This is what Gräper does in his notion of dance, which lucidly illustrates the definition of movements proposed by process philosophy, namely, as "connected developments" and "orchestrated series of occurrences" that unfold in "programmatic coordination"—each alone, yet bound up with the others "either causally or functionally."[84] Unlike objects, events and movements do not attain connection, continuity, and structure through demarcation from something external to themselves, but by producing themselves as an ensemble through internal interreference. Rules determine and coordinate how movements relate to one another and succeed one another. In that process, each movement can be constituted only of further movements and only in the context of other movements.

At the nanoscopic level, everything is movement. Movements here are not fixed events; they exist only in "reciprocal interaction."[85] Mathematical operations take the data sets garnered from microscopy and—using modeling, differential equations, numerical analysis, or statistical procedures—integrate them into nanoscopic movements that may therefore be regarded as stability patterns or statistical regularities. These methods enable motion to be aggregated on different levels and to become manifest on each different level of integration as an event in time and space. In each case, movement is described in the form of a system of rules. Yet molecular motion research has made proteins active players in movement. Kinesins, myosins, and microtubules are described in terms of things that exhibit agency in their motion—as motors that "do" life.

Process philosophy offers a way to resolve this tension between rules and individual agency by changing the perspective. Setting events above things, the process approach reverses their roles: "processes are not the machinations of stable things; things are the stability patterns of variable processes."[86] Is it, then, perhaps no more than a change of perspective that lies between the kinesin walker as a concrete agent of life and as the expression of a stable pattern of mathematical formalism—a switch from thing to process, from a kind of vitalism to a kind of mechanism, from animation to formula? Does this switch open up a path to understanding the "unheard-of event," in Goethe's words, that occurred when the production of movement in a test tube led to animations of walking molecules?

In fact, the two ways of describing molecular movement cannot be reconciled through a shift in perspective; they squarely collide. At the molecular level, mathematical formalization calculates the most probable movement among a set of possible movements. *That* movement happens in molecules, then, can become apparent solely through statistics and probability. But what the kinesin walker, for example, does is to highlight *how* movement happens, by attributing specific ways of moving to molecules. And asking how movement happens takes us beyond statistics and probability, into the realm of the possible and the potential.

As an animal feature, and especially a human feature, walking is characterized by its potentiality. In walking, what may strike us as most intriguing is less local displacement than, to quote the philosopher of art Erin Manning, the "incipiency" of movement, the "qualitative multiplicity" it contains—a potential that is always there, but will "always remain not-yet." Animal movement contains the potential of change from that which we expect: the practice and repetition of steps is always "opening toward different shades of movement."[87] When observing a pedestrian's aimless stroll or an animal's purposeful hunt, for example, at no point in time can we predict what will happen next. We may reason that the pedestrian will most probably simply continue to stroll, the predator continue to chase its prey, but at any moment, without warning, they may switch into another mode of movement. This quality of animal movement as a transition from a life that is potential to an actualized life was at the core of the Aristotelian view of motion. There, being real meant being-in-the-world and being-on-the-way-to-completion, a movement that was actively directed at self-maintenance and the individual's role as part of a greater whole. The potentiality inherent in animal motion—a quality no probabilistic pattern can express and no motor incarnate—thus ensured life's continuity and identity as it moved toward the future.

In this book, we have followed motion on its way from the soul, human beings, and other animals via single-celled organisms and cells to protoplasm and proteins. From Leeuwenhoek, via Lamarck, to Cohn and Gräper, observers struggled to understand how it could be that motion appeared wherever they looked. As they shifted their gaze from wholes to parts, life remained beyond dispute: life was what brought about motion wherever motion was found. Today, it is motion that is beyond dispute. Despite its technological constitution and because of its mathematical formulability, motion's ubiquity is generally acknowledged; it no longer needs to be specially affirmed. However, doubts are arising as to whether motion is alive.

Contemporary biology, by increasingly undertaking to produce life, increasingly blurs the frontier between nature-become-art and artificiality-become-nature. In this hybrid living world, biology can uphold its scientific enterprise only if it can show that the life it produces is, indeed, alive. For however much biology has changed over the centuries, one thing has remained constant: motion is still regarded as key to life. It is the particular way in which matter moves that persuades us to credit it with life.

For the new world of biology, therefore, animate motion is even more significant than before. It is no longer the signum only of the living as we find it, but also that of a world made to be alive. Yet the movement made apparent by the mathematical tools and formalisms of present-day biology becomes apparent solely through aggregated statistical and probabilistic data. Movement generated in this way does not have the qualities of life. When soft robots creep and twine through the human body, molecular machines walk through the cytoplasm, or cells dance through the ovum, we attribute vitality to them not by noting their statistical patterns, but by ascribing to them the very qualities in which Aristotle once located life. Twenty-first-century biology needs those qualities more than ever before in order to justify its ventures in the making of new living worlds. As a result, the mathematization of motion—that Cartesian apotheosis of mechanization—ultimately fails to free itself of the Aristotelian qualities it set out to eliminate once and for all.

Epilogue

Seeking to capture past centuries' thinking on life, the history and theory of biology can make use of many different perspectives. Up to now, motion has not been among them. Yet my concern is not to follow the academic zeitgeist by proclaiming some unheard-of advance. The journey of this book was guided solely by curiosity — the attraction exerted by an endlessly fascinating object. My aim was nothing more and nothing less than to consider the past of biology through the prism of motion.

Looking at motion, this book looks at something that is taken for granted. Motion lives in a space of ambivalence: it is there, but always already over; ubiquitously present to attention, but elusive to cognition. Its very ubiquity seems to make it the perfect Ariadne's thread along which to feel our way through the centuries, crosscutting scientific enclosures and ignoring their claims to interpretive sovereignty. Following the thread, we pass through terrain that has long since been scoured for meaning, measured, and thought through, but that now becomes alien, awkward again — only to reveal itself on the way back as entirely familiar after all.

That is true not least for two conceptual structures that, in their various iterations, have stamped the study of living matter from the start: mechanism and vitalism. Like all dichotomies, they prove difficult to sustain, and their object, biology, anyway resists simple binary divisions. But contemplating biology through notions of movement cannot mean accusing the conceptual pair of inadequacy, resolving their contradiction or recasting it as a necessary complementarity, or propounding motion as a brand-new solution. Rather, an observer pursuing this path of motion, from ensouled animals to molecular motors, will simply arrive at a different destination from those of previous routes.

Having reached that destination, mechanism and vitalism turn out to be equally at sea. Nowadays, biological motion no longer "is," it is made. Things that move as only a living body used to do are no longer indubitably alive. For centuries, motion was the lowest common denominator and the basic feature of all life, its ineluctable foundation; today biologists model it mathematically or instantiate it experimentally with magnetic fields, inscribing what they have made with the aliveness of movement.

What I have tried to show here is that this inscription is not a random act. Technical and methodological developments, the toolbox of mechanical procedure and formal analysis, have led biology to this provisional culmination of the achievable. Biological motion, however, sets limits on what biology can do. On the one hand, these lie in the multiplicity of ways in which the living world is able to move, a multiplicity that is key to the orchestration of motion happening in the body's interior. On the other, they lie in the individual. As early as the seventeenth century, the diversity revealed by studying dead organisms led anatomists to doubt mechanism's central assumption of always-identical reproduction and always-identical functioning.

Even more than the dead body, and unlike the organism of biotechnological synthesis, the living body and its movements never cease to be contingent. They are the mover's individual projection into the future, a step on a path that could be different at every turn. Without that contingency, motion would not be able to secure the continued existence of every single organism. The space of motion — always potentially, spontaneously different — marks out the two existential presuppositions of life: its continuity and its identity as it moves forward.

It is this promise of continuation and orientation toward the future that forms the core of vitalism. To be sure, vitalism could never do more than postulate it, but it was built upon the experience that biological life continues of its own accord, at least wherever it is not specifically prevented from doing so. What is new in the present day is, first, that molecular motion now lies at the heart of mechanism, and second, that the living world's inherent promise for the future has taken on a questionable hue in the context of the Anthropocene.

The science of life, which always banished vitalism to the nonscientific realm, has thus reached a point where it must itself take ownership of this promise for the future. Biology ascribes living motion through its methods and experiments, describing swimming sperm, rolling leukocytes, and limping motors while simultaneously calculating probabilistic patterns. It signals that life sustains itself into the future, as the Aristotelian quality of being-in-the-world and being-on-the-way-to-completion, by cleaving to the most ancient notion of aliveness despite the fact that no probabilistic pattern and no motor displays that quality.

In contrast, the work of Ludwig Gräper, Lewis Wolpert, and Tryggve Gustafson or thinking in process philosophy hints that the Ariadne's thread available to us — however it may ultimately turn out to be constituted — is precisely only a way back. Perhaps that is what Gregor Samsa realizes when he turns around to die, for the first time crawling into his room forward instead of backward. Finding himself on an unfamiliar path, he has no choice but to move ahead unknowing.

Acknowledgments

This book's journey began at the Wissenschaftskolleg zu Berlin in the leafy district of Grunewald. More precisely, it started with Graham Greene's *Travels with My Aunt*, on those charming, late summer days in 2013 when reading about Henry Pulling and writing about biological motion came together in a happy first companionship. I cannot exactly claim with Greene that this was the only book that was fun to write. All the less so because I have not written twenty-seven novels to make a fair comparison and because the cast of *Biological Motion* does not dazzle with dahlias, smugglers, and eccentric aunts. Nor was it written in France, for that matter. But in the stillness of a global pandemic, into which my writing would soon head, Aunt Augusta continued to inspire in me a sense of waywardness and curiosity. When Henry Pulling is "weeding the dahlias, the Polar Beauties and the Golden Leaders and the Requiems," the telephone suddenly begins to ring, shattering the quiet of his little garden.

I want to thank everyone at and around the Wissenschaftskolleg for making my time there so rich and enabling so many ideas for the book, ideas of which I became aware only much later. I also spent an important year at the Radcliffe Institute for Advanced Study at Harvard University, housed in Brattle Street and nurtured by the wonderful friendship of Shireen Hassim, Camilo Mendez, and Silvija Dovydenaite, now all far too far away. I enormously enjoyed their intellectual and caring company, as well as discussions with my fellow fellows and the support from all the institute's members and staff. I thank Eleanor Bragg and David Xiang for their work and enthusiasm as Radcliffe Research Partners during that year and the Harris Program, especially the wonderful Jane E. Lipson, for inviting me to spend a beautiful spring at Dartmouth in 2022. Leuphana University Lüneburg — and Claus Pias at the university's Research Institute for Media Cultures of Computer Simulation — made possible all of my and the book's formative travels during the years of my employment there, for which I am very grateful.

It is a great privilege to have my second monograph published with Zone Books. Ramona Naddaff and Meighan Gale have a gift for making intellectual journeys metamorphose into a visual and haptic delight. My warmest thanks to them and to Shigehisa Kuriyama, Hannah Landecker, Friedrich Steinle, Christina Brandt, and Arianna Borrelli for all their valuable comments and insights.

I am happy to confess, finally, that for Kate Sturge, *Biological Motion* was not so much a book as a test of her mettle. Written partly in German, partly in English, it took shape between languages — a most exciting, frustrating, and intellectually rewarding experience. Words teach so much. I am infinitely grateful to Kate for her trust and sensitivity and for her willingness to devote her astute and poetic mind to this book.

Notes

Translator's note: throughout, translations of quoted material are my own unless otherwise attributed.

PROLOGUE

1. All quotations are from Franz Kafka, "The Metamorphosis," trans. Willa and Edwin Muir, in *The Complete Stories*, ed. Nahum N. Glatzer (New York: Schocken, 1971), pp. 89–139.

INTRODUCTION

1. On the human body in Kafka's work, see Robert Sell, *Bewegung und Beugung des Sinns: Zur Poetologie des menschlichen Körpers in den Romanen Franz Kafkas* (Weimar: Metzler, 2002). On motion as an expressive faculty specific to the human body in Kafka's letters and journals, see Elisabeth Lack, *Kafkas bewegte Körper: Die Tagebücher und Briefe als Laboratorien von Bewegung* (Paderborn: Fink, 2009). For most interpreters of Kafka's "Metamorphosis," the repertoire of human and animal movements serves as a ground upon which Kafka can explore his real concern: a reflection on human existence. The story has lost none of its enigmatic allure today. Vladimir Nabokov, for example, an entomologist especially interested in butterflies, carefully considered the identification of Kafka's *Ungeziefer* ("vermin") as a beetle, but despite admiring Kafka's precision, he read the story as a parable on art. See Leland De La Durantaye, "Kafka's Reality and Nabokov's Fantasy: On Dwarves, Saints, Beetles, Symbolism, and Genius," *Comparative Literature* 59.4 (2007), pp. 315–31.

2. Nicholas Rescher, *Process Philosophy: A Survey of Basic Issues* (Pittsburgh, PA: University of Pittsburgh Press, 2000), pp. 77 and 76. *Merriam-Webster's Dictionary* (www.

merriam-webster.com/dictionary/artifice) defines "artifice" as "clever or artful skill," "artful stratagem," or "false or insincere behavior."

3. On the origins of European philosophy in the opposition between Parmenides and Heraclitus, see Margot Fleischer, *Anfänge europäischen Philosophierens: Heraklit, Parmenides, Platons Timaios* (Würzburg: Königshausen & Neumann, 2001); André Laks and Claire Louguet, eds., *Qu'est-ce que la philosophie présocratique? What Is Presocratic Philosophy?* (Villeneuve d'Ascq: Presses Universitaires de Septentrion, 2002); André Laks, *The Concept of Presocratic Philosophy: Its Origin, Development, and Significance*, trans. Glenn W. Most (Princeton, NJ: Princeton University Press, 2018).

4. Friedrich Kaulbach and Gerbert Meyer, s.v. "Bewegung," in *Historisches Wörterbuch der Philosophie*, ed. Joachim Ritter (Basel: Schwabe, 1971).

5. The philosophical study of motion is marked by a rupture between ancient philosophy, which thought about motion in qualitative terms, negotiating it through oppositions such as unity/multiplicity, mutability/immutability, or being/nonbeing, and a modern, scientific description of the laws of motion using physical and mathematical concepts. See Michel Blay, *La science du mouvement: De Galilée à Lagrange* (Paris: Belin, 2002). In Kant's transcendental philosophy, motion carries little weight compared with the categories of time and space. See Friedrich Kaulbach, *Der philosophische Begriff der Bewegung: Studien zu Aristoteles, Leibniz und Kant* (Cologne: Böhlau, 1965). In Henri Bergson's philosophy, the crucial experience of the reality of process and motion is located in the concept of *durée* — not the time of physics, but the time of the living world, a fundamental datum of consciousness — and can be comprehended only through intuition; see Bergson, *The Creative Mind: An Introduction to Metaphysics*, trans. Mabelle L. Andison (New York: Philosophical Press, 1946). In *Creative Evolution*, too, at the beginning of the twentieth century, Bergson finds that life, evolution, and nature's creativity do not seem comprehensible by mechanistic or teleological principles, and he therefore adduces an (unexplained) *élan vital* to describe nature. Bergson, *Creative Evolution*, trans. Arthur Mitchell (New York: Henry Holt, 1911). In the twentieth century, the processuality and constant becoming of the living world took up a central place in the philosophies of Alfred North Whitehead and Gilles Deleuze. Whitehead, *Process and Reality: An Essay in Cosmology* (Cambridge: Cambridge University Press, 1929); Deleuze, *Bergsonism*, trans. Hugh Tomlinson and Barbara Habberjam (New York: Zone Books, 1991).

6. Lorraine Daston, "The History of Science and the History of Knowledge," *KNOW: A Journal on the Formation of Knowledge* 1.1 (2017), p. 136.

7. The "snail-o-bot" is part of the research in Metin Sitti's lab at the Max Planck Institute for Intelligent Systems. The lab's previous work on a soft robot that moves, but is not alive, is discussed in Chapter 4. See Hamed Shahsavan et al., "Bioinspired Underwater Locomotion of Light-Driven Liquid Crystal Gels," *Proceedings of the National Academy of Sciences* 117.10 (2020), pp. 5125–33. Computation, robotics, and integrative approaches to the interaction of motion with neurological and other systems and with the environment have boosted research in this field in recent decades. For a survey, see Michael H. Dickinson et al., "How Animals Move: An Integrative View," *Science* 288.5463 (2000), pp. 100–106. These perspectives also cover the motion of bacteria, viruses, or sperm using flagella, cilia, or pseudopodia; see Chapter 4.

8. *Encyclopaedia Britannica*, s.v. "animal behaviour," www.britannica.com/science/animal-behavior.

9. Marcel Mauss, "Techniques of the Body" (1934), trans. Ben Brewster, *Economy and Society* 2.1 (1973), p. 70. See Jean-François Loudcher, "Limites et perspectives de la notion de Technique du Corps de Marcel Mauss dans le domaine du sport," *STAPS* 91.1 (2011), pp. 9–27. In the French tradition after Mauss, the relevant approaches are, first, based on the history of technology and culture, which can be traced from Mauss's student, the palaeontologist and palaeoanthropologist André Leroi-Gourhan (1911–1986), up to the present day, for example in the philosophy of Bernard Stiegler (1952–2020), and second, the more broadly conceived French history of the body, such as the three-volume *Histoire du corps* edited by Alain Corbin, Jean-Jacques Courtine, and Georges Vigarello (Paris: Seuil, 2005–2006).

10. Of the abundant literature, see, for example, on upright walking, Hans Blumenberg, *Beschreibung des Menschen*, ed. Manfred Sommer (Frankfurt am Main: Suhrkamp, 2006), and Kurt Bayertz, *Der aufrechte Gang: Eine Geschichte des anthropologischen Denkens* (Munich: C. H. Beck, 2012). On the flaneur, see Honoré de Balzac, *Comédie humaine*, 7 vols. (New York: Century, 1911), and Walter Benjamin, *The Arcades Project*, trans. Howard Eiland and Kevin McLaughlin (Cambridge, MA: Belknap Press of Harvard University Press, 1999). On walking more generally, see Andreas Mayer, *The Science of Walking: Investigations into Locomotion in the Long Nineteenth Century*, trans. Tilman Skowroneck and Robin Blanton (Chicago: University of Chicago Press, 2020). On recent work in the history and anthropology of dance, see Geraldine Morris and Larraine Nicholas, eds., *Rethinking Dance History: Issues and Methodologies* (London: Routledge, 2018); Hélène Neveu Kringelbach and Jonathan Skinner, eds., *Dancing Cultures: Globalization, Tourism*

and Identity in the Anthropology of Dance (New York: Berghahn, 2014), and Drid Williams, *Anthropology and the Dance: Ten Lectures* (Urbana: University of Illinois Press, 2004). On body, movement, and perception around 1900, see Stephen Kern, *The Culture of Time and Space, 1880–1918* (Cambridge, MA: Harvard University Press, 1983).

11. The scholarship in this area is very wide-ranging. See Anson Rabinbach, *The Human Motor: Energy, Fatigue, and the Origins of Modernity* (Berkeley: University of California Press, 1992). For European history and sports culture, see Horst Bredekamp, *Florentiner Fußball: Die Renaissance der Spiele* (Berlin: Wagenbach, 2001), and Julia Allen, *Swimming with Dr. Johnson and Mrs. Thrale: Sport, Health and Exercise in Eighteenth-Century England* (Cambridge: Lutterworth Press, 2012). For a survey of the field, see Rebekka von Mallinckrodt and Angela Schattner, eds., *Sports and Physical Exercise in Early Modern Culture: New Perspectives on the History of Sports and Motion* (Abingdon: Routledge, 2016), and Rebekka von Mallinckrodt, ed., *Bewegtes Leben: Körpertechniken in der Frühen Neuzeit* (Wolfenbüttel: Herzog August Bibliothek, 2008). A classic in the sociology of sport is Norbert Elias and Eric Dunning, eds., *Quest for Excitement: Sport and Leisure in the Civilizing Process* (Oxford: Blackwell, 1986). On prisons, see Michel Foucault, *Discipline and Punish: The Birth of the Prison*, trans. Alan Sheridan (New York: Vintage Books, 1979).

12. On physiognomy, see Simon Swain, ed., *Seeing the Face, Seeing the Soul: Polemon's Physiognomy from Classical Antiquity to Medieval Islam* (Oxford: Oxford University Press, 2007), and Martin Porter, *Windows of the Soul: The Art of Physiognomy in European Culture 1470–1780* (Oxford: Oxford University Press, 2005). On theater, see Claudia Jeschke and Hans-Peter Bayerdörfer, eds., *Bewegung im Blick: Beiträge zur theaterwissenschaftlichen Bewegungsforschung* (Berlin: Vorwerk 8, 2000). On baroque spectacle, see Nicola Gess, Tina Hartmann, and Dominika Hens, eds., *Barocktheater als Spektakel: Maschine, Blick und Bewegung auf der Opernbühne des Ancien Régime* (Paderborn: Fink, 2015). On social perception, see M. D. Rutherford and Valerie A. Kuhlmeier, eds., *Social Perception: Detection and Interpretation of Animacy, Agency, and Intention* (Cambridge, MA: MIT Press, 2013), and Chapter 3 of this book.

13. A general introduction is offered by Andreas Beyer and Guillaume Cassegrain, eds., *Mouvement/Bewegung: Über die dynamischen Potenziale der Kunst* (Berlin: Deutscher Kunstverlag, 2015). On architecture, see Spyros Papapetros, *On the Animation of the Inorganic: Art, Architecture, and the Extension of Life* (Chicago: University of Chicago Press, 2012). Still foundational are Aby Warburg's discussions of movement and moving accessories in Florentine Renaissance art: Warburg, *Werke in einem Band*, ed. Martin Treml,

Sigrid Weigel, and Perdita Ladwig (Frankfurt am Main: Suhrkamp, 2010). Art historian David Freedberg's studies on psychological responses in art have recently embraced neurophysiology, for example, Freedberg and Vittorio Gallese, "Motion, Emotion and Empathy in Esthetic Experience," *Trends in Cognitive Sciences* 11.5 (2007), pp. 197–203.

14. Quoted in Martin Kemp, ed., *Leonardo on Painting* (New Haven, CT: Yale University Press, 2001), p. 144. Even today, Leonardo's explanations, sketches, and paintings remain a touchstone in the visual appropriation of motion and the transformation of movement — whether the movement of humans or other animals, rivers, or vortices — into the impression of life in a picture surface. See Alexander Perrig, "Leonardo: Die Rekonstruktion menschlicher Bewegung," in Gottfried Schramm, ed., *Leonardo: Bewegung und Ruhe* (Freiburg: Rombach, 1997), pp. 67–100; Frank Fehrenbach, *Leonardo da Vinci: Der Impetus der Bilder* (Berlin: Matthes & Seitz, 2019), On the topos of liveliness in early modern art, see Fehrenbach, *Quasi vivo: Lebendigkeit in der italienischen Kunst der Frühen Neuzeit* (Berlin: De Gruyter, 2020).

15. Martin Kemp, "Die Zeichen lesen: Zur graphischen Darstellung von physischer und mentaler Bewegung in den Manuskripten Leonardos," in Frank Fehrenbach, ed., *Leonardo da Vinci: Natur im Übergang* (Munich: Wilhelm Fink, 2002), p. 226 (original emphasis).

16. Ibid., p. 207.

17. On scientific cinematography, see Chapter 7; on animation, see Chapter 8 of this book. On molecular animation and digital entertainment, see Adam Nocek, *Molecular Capture: The Animation of Biology* (Minneapolis: University of Minnesota Press, 2021). On dance notation, see Whitney E. Laemmli, "The Living Record: Alan Lomax and the World Archive of Movement," *History of the Human Sciences* 31.5 (2018), pp. 23–51, and Laemmli, "Paper Dancers: Art as Information in Twentieth-Century America," *Information & Culture* 52.1 (2017), pp. 1–30. On the history and technologies of representation from notation to motion capture, see Nicolás Salazar Sutil, *Motion and Representation: The Language of Human Movement* (Cambridge, MA: MIT Press, 2015); Hubertus von Amelunxen, Dieter Appelt, and Michael Baumgartner, eds., *Notation: Kalkül und Form in den Künsten* (Berlin: Akademie der Künste Berlin, 2008); Pirkko Rathgeber, "Struktur- und Umriß-modelle als schematische Bilder der Bewegung," *Rheinsprung 11: Zeitschrift für Bildkritik* 2 (2011), pp. 130–53; Sigrid Leyssen and Pirkko Rathgeber, eds., *Bilder animierter Bewegung/ Images of Animate Movement* (Munich: Fink, 2013).

18. Lisa Chong, Elizabeth Culotta, and Andrew Sugden, "On the Move," *Science* 288.5463 (2000), p. 79.

19. Kevin Jiang, "To Boldly Go: New Microscope Captures 3-D Movies of Cells inside Living Organisms in Unprecedented Detail," Harvard Medical School News & Research, April 19, 2018, https://hms.harvard.edu/news/boldly-go.

20. On this discussion in the arts, different aesthetic and artistic attitudes to siting, and postmodern developments, see Peter Gillgren, *Siting Michelangelo: Spectatorship, Site Specificity and Soundscape* (Lund: Nordic Academic Press, 2017), pp. 30 and 29. Gillgren also notes that "actual sensory experience" is surprisingly absent from art history's discussion of siting, which is "overtly logocentric" (p. 29).

21. For an impression of how dramatic the new experience of site-specific art seemed to viewers in the 1960s, see Douglas Crimp, "Serra's Public Sculpture: Redefining Site Specificity," in Laura Rosenstock, ed., *Richard Serra: Sculpture* (New York: The Museum of Modern Art, 1986), pp. 40–56; for an overview, see Miwon Kwon, *One Place after Another: Site-Specific Art and Local Identity* (Cambridge, MA: MIT Press, 2002).

22. See Alexander Nagel, *Medieval Modern: Art out of Time* (London: Thames & Hudson, 2012); Gillgren, *Siting Michelangelo.*

23. Gillgren, *Siting Michelangelo*, pp. 29–30.

24. Rescher, *Process Philosophy*, p. 12.

25. Gillgren, *Siting Michelangelo*, p. 29.

26. Kemp, "Die Zeichen lesen," p. 226.

CHAPTER ONE: WHAT ITSELF MOVES ITSELF

1. Aristotle, *Physics*, trans. R. P. Hardie and R. K. Gaye, in *The Basic Works of Aristotle*, ed. with an introduction by Richard McKeon (New York: Random House, 1941), p. 236.

2. This is the translation chosen by Hardie and Gaye in *The Basic Works of Aristotle*, pp. 253–54. There are various other English translations of the phrase, for example, "the actuality of a potentiality as such" (Joe Sachs, "Aristotle: Motion," *Internet Encyclopedia of Philosophy* [2005, www.iep.utm.edu/aris-mot], ch. 2); "the coming to be actual of the potential, insofar as it is potential" (Christopher Byrne, *Aristotle's Science of Matter and Motion* [Toronto: University of Toronto Press, 2018], p. 15); and "the realizing of a potentiality, *qua* potentiality" (Aristotle, *The Physics*, trans. Francis M. Cornford and Philip H. Wicksteed, 2 vols. [Cambridge, MA: Harvard University Press, 1980], vol. 1, p. 195).

3. Sachs, "Aristotle," ch. 2. As Friedrich Kaulbach puts it: "Something that is, stands at the tipping point from one state to the other." Kaulbach reads entelechy as meaning "the whole in the process of achieving its actuality." Kaulbach, *Der philosophische Begriff*

der Bewegung: Studien zu Aristoteles, Leibniz und Kant (Cologne: Böhlau, 1965), pp. 4 and 5. He adds (p. 7): "Achieving the whole of motion, in which all momentum is aggregated, is the *energeia* that is simultaneously also *entelecheia*, that is, what embraces the beginning, middle, and end. Entelechy is identical for mover and movable, for *dunamis* and *energeia*."

4. Sachs reads *entelecheia* as "continuing in a state of completeness," and actuality as "to be at work" and "to act for an end." Post-Aristotelian philosophy wrestled with the opposition between actuality and potentiality. St. Thomas Aquinas famously reconciled it "by arguing that in every motion actuality and potentiality are mixed or blended"; Descartes did so simply by interpreting motion as the passage from potentiality to actuality. Sachs, "Aristotle."

5. Sachs, "Aristotle"; Friedrich Kaulbach and Gerbert Meyer, s.v. "Bewegung," in Joachim Ritter, ed., *Historisches Wörterbuch der Philosophie* (Basel: Schwabe, 1971); Kaulbach, *Der philosophische Begriff der Bewegung*.

6. Aristotle, *De anima*, trans. J. A. Smith, in *The Basic Works of Aristotle*, ed. with an introduction by Richard McKeon (New York: Random House, 1941), p. 535. See also *DA* 2.1.412a13: "Of natural bodies some have life in them, others not; by life we mean self-nutrition and growth (with its correlative decay)."

7. Aristotle, *Parva Naturalia with On the Motion of Animals*, trans. David Bolotin (Macon, GA: Mercer University Press, 2021), p. vii.

8. Aristotle, *De motu animalium*, trans. A. S. L. Farquharson, in *The Works of Aristotle*, eds. J. A. Smith and W. D. Ross, vol. 5 (Oxford: Clarendon Press of Oxford University Press, 1912).

9. A newly reconstructed text of *De motu animalium* can be found in a recent critical German edition: Aristoteles, *De motu animalium: Über die Bewegung der Lebewesen, Griechisch-Deutsch*, ed. Oliver Primavesi, trans. Klaus Corcilius (Hamburg: Felix Meiner, 2018), pp. clxvii–clxxxii, and in the new English edition by Bolotin, *Parva Naturalia with On the Motion of Animals*, pp. 143–76. These recent editions are part of a renewed interest in Aristotle's biology in general and his theory of animal motion in particular, an interest whose foundations were laid by Allan Gotthelf and James G. Lennox, eds., *Philosophical Issues in Aristotle's Biology* (Cambridge: Cambridge University Press, 1987). See also James G. Lennox, *Aristotle's Philosophy of Biology: Studies in the Origins of Life Science* (Cambridge: Cambridge University Press, 2001); Lennox, "The Complexity of Aristotle's Study of Animals," in Christopher Shields, ed., *The Oxford Handbook of Aristotle* (Oxford: Oxford University Press, 2012), pp. 287–305. Other recent works include Pierre-Marie Morel, *De*

la matière à l'action: Aristote et le problème du vivant (Paris: Vrin, 2007); Jean-Louis Labarrière, *Langage, vie politique et mouvement des animaux: Études aristotéliciennes* (Paris: Vrin, 2004); André Laks and Marwan Rashed, eds., *Aristote et le mouvement des animaux: Dix études sur le* De motu animalium (Villeneuve d'Ascq: Presses Universitaires de Septentrion, 2004); and Jason A. Tipton, *Philosophical Biology in Aristotle's Parts of Animals* (Heidelberg: Springer, 2014). Tipton aligns Aristotle's description of particular movements with particular zoological species.

10. See the extensive discussion in Aristoteles, *De motu animalium*, notes pp. 163–69.

11. See Klaus Corcilius and Pavel Gregoric, "Aristotle's Model of Animal Motion," *Phronesis* 58.1 (2013), pp. 52–97. Corcilius and Gregoric discuss problems with, for example, the mechanical form in which Aristotle, having no concept of the central nervous system, envisioned the transmission and amplification of impulses in the body (from the heart to the periphery and vice versa). For more on the automaton model, see Aristoteles, *De motu animalium*, pp. cx–cxxvi and 144–48. On Aristotle's mechanical knowledge in general, and of automata and circular movements in particular, see also Jean De Groot, *Aristotle's Empiricism: Experience and Mechanics in the Fourth Century BC* (Las Vegas, NV: Parmenides, 2014), and Jean De Groot, "*Dunamis* and the Science of Mechanics: Aristotle on Animal Motion," *Journal of the History of Philosophy* 46.1 (2008), pp. 43–67.

12. Aristotle's ideas on animal movement were ignored by mechanist approaches because he failed to recognize the role of muscles in the execution of movement. See the summary, with a detailed discussion of the concepts, in Pavel Gregoric and Martin Kuhar, "Aristotle's Physiology of Animal Motion: On *Neura* and Muscles," *Apeiron* 47.1 (2014), pp. 94–115; also Corcilius and Gregoric, "Aristotle's Model of Animal Motion," and Michael Frampton, *Embodiments of Will: Anatomical and Physiological Theories of Voluntary Animal Motion from Greek Antiquity to the Latin Middle Ages, 400 B.C.–A.D. 1300* (Saarbrücken: VDM, 2008), pp. 85–101.

13. Corcilius and Gregoric, "Aristotle's Model of Animal Motion," p. 98. This accords with the general approach of *Physics* and its insistence that all movements in nature can be derived from a first, unmoved mover.

14. See the presentation of fundamental principles in *De anima* and *De motu animalium*. *De incessu animalium* examines the various mechanisms of motion in the external organs of movement, and there are further references to specific parts of the body in *Historia animalium* and *De partibus animalium*. Corcilius studies the writings as a single systematic unit and as proposing a theory of locomotion that applies to all living beings.

Klaus Corcilius, *Streben und Bewegen: Aristoteles' Theorie der animalischen Ortsbewegung* (Berlin: De Gruyter, 2008); Gregoric and Kuhar, "Aristotle's Physiology."

15. The issue of perception and intentionality is among the most difficult and controversial aspects of Aristotle's theory of animal motion. See three essays in Mary Louise Gill and James G. Lennox, eds., *Self-Motion: From Aristotle to Newton* (Princeton, NJ: Princeton University Press, 1994): David J. Furley, "Self-Movers," Mary Louise Gill, "Aristotle on Self-Motion," and Cynthia A. Freeland, "Aristotle on Perception, Appetition, and Self-Motion." Also Martha Craven Nussbaum, *Aristotle's De motu animalium: Text with Translation, Commentary, and Interpretive Essays* (Princeton, NJ: Princeton University Press, 1985). On rational action as a particular form of animal locomotion, see Patricio A. Fernandez, "Reasoning and the Unity of Aristotle's Account of Animal Motion," in *Oxford Studies in Ancient Philosophy*, vol. 47, ed. Brad Inwood (Oxford: Oxford University Press, 2014). Corcilius proposes a new interpretation of Aristotle's approach as a "theory of striving" (*Theorie der Strebung*): Corcilius, *Streben und Bewegen*, part 1, and Aristoteles, *De motu animalium*, pp. ccii–ccxl. He concludes that for Aristotle, self-motion is not a distinct faculty of the soul, but a "psychophysical process." Aristoteles, *De motu animalium*, p. clxxx.

16. On the unmoved mover, see Byrne, *Aristotle's Science*, p. 19.

17. Thus Corcilius in his introduction to Aristoteles, *De motu animalium*, p. cli.

18. Wolfgang Wieland, *Die aristotelische Physik* (Göttingen: Vandenhoeck & Ruprecht, 1962), pp. 231–54, esp. p. 245. Corcilius resolves the dilemma by excepting from animal autonomy the energy necessary for movement, for which the animal depends upon its environment. Aristoteles, *De motu animalium*, p. cc.

19. For details, see Wieland, *Die aristotelische Physik*, pp. 248–49.

20. Hans Blumenberg, *Paradigms for a Metaphorology*, trans. Robert Savage (Ithaca, NY: Cornell University Press, [1960] 2010), p. 63. On ancient automata, see Pascal Weitmann, *Technik als Kunst: Automaten in der griechisch-römischen Antike und deren Rezeption in der frühen Neuzeit als Ideal der Kunst oder Modell für Philosophie und Wissenschaft* (Tübingen: Wasmuth, 2011), esp. p. 8 on Egyptian dolls; Alfred Espinas, "L'organisation ou la machine vivante en Grèce, au IVe siècle avant J.-C.," *Revue de Métaphysique et de Morale* 11.6 (1903), pp. 703–15; Sylvia Berryman, "The Imitation of Life in Ancient Greek Philosophy," in Jessica Riskin, ed., *Genesis Redux: Essays in the History and Philosophy of Artificial Life* (Chicago: University of Chicago Press, 2007), pp. 35–45.

21. On medieval automata from the ninth to the fifteenth century, see E. R. Truitt,

Medieval Robots: Mechanics, Magic, Nature, and Art (Philadelphia: University of Pennsylvania Press, 2015); on the chronology, pp. 2–8. Jessica Riskin unravels the epistemic complexity of the automaton's agency and its metaphysical mediation between the mortal world and the afterlife: Jessica Riskin, *The Restless Clock: A History of the Centuries-Long Argument over What Makes Living Things Tick* (Chicago: University of Chicago Press, 2016), pp. 11–43; see also Jessica Keating, *Animating Empire: Automata, the Holy Roman Empire, and the Early Modern World* (University Park: Pennsylvania State University Press, 2018). On eighteenth-century androids, see Adelheid Voskuhl, *Androids in the Enlightenment: Mechanics, Artisans, and Cultures of the Self* (Chicago: University of Chicago Press, 2013); from a historical and epistemological perspective more generally, see Minsoo Kang, *Sublime Dreams of Living Machines: The Automaton in the European Imagination* (Cambridge, MA: Harvard University Press, 2011), pp. 79–80; Jessica Riskin, ed., *Genesis Redux: Essays in the History and Philosophy of Artificial Life* (Chicago: University of Chicago Press, 2007); and Aurélia Gaillard et al., eds., *L'automate: Modèle, métaphore, machine, merveille* (Pessac: Presses universitaires de Bordeaux, 2013).

22. Georges Canguilhem, *Knowledge of Life*, trans. Stefanos Geroulanos and Daniela Ginsburg (New York: Fordham University Press, 2008), pp. 76–77.

23. Luisa Dolza and Hélène Vérin, "Figurer la mécanique: L'énigme des théâtres de machines de la Renaissance," *Revue d'histoire moderne et contemporaine* 51.2 (2004), pp. 7–37; Jonathan Sawday, *Engines of the Imagination: Renaissance Culture and the Rise of the Machine* (London: Routledge, 2007); John R. Pannabecker, "Representing Mechanical Arts in Diderot's *Encyclopédie*," *Technology and Culture* 39.1 (1998), pp. 33–73.

24. Ofer Gal and Yi Zheng, eds., *Motion and Knowledge in the Changing Early Modern World: Orbits, Routes and Vessels* (Dordrecht: Springer, 2014); Pamela H. Smith and Paula Findlen, eds., *Merchants and Marvels: Commerce, Science, and Art in Early Modern Europe* (London: Routledge, 2002).

25. On the last of these points, see Matthew L. Jones, *Reckoning with Matter: Calculating Machines, Innovating, and Thinking about Thinking from Pascal to Babbage* (Chicago: University of Chicago Press, 2016).

26. Alexander Sutter, *Göttliche Maschinen: Die Automaten für Lebendiges bei Descartes, Leibniz, La Mettrie und Kant* (Frankfurt am Main: Athenäum, 1988), p. 24.

27. On the idea that Cartesian reason is not part of nature, but is required by nature, see Kaulbach, *Der philosophische Begriff der Bewegung*, pp. 31–32.

28. See, for example, the cosmologies of Thomas Hobbes (1588–1679) and Pierre

Gassendi (1592–1655). They are based on motion caused by tiny, physical particles or atoms, which are dynamic and mobile in themselves. For a discussion of Gassendi's atomist matter theory with attention to his notion of movement and his reading of Hobbes, see Lisa T. Sarasohn, "Motion and Morality: Pierre Gassendi, Thomas Hobbes and the Mechanical World-View," *Journal of the History of Ideas* 46.3 (1985), pp. 363–79; Frithiof Brandt, *Thomas Hobbes' Mechanical Conception of Nature* (Copenhagen: Levon & Munksgaard, 1927).

29. See Dennis Des Chene, *Spirits and Clocks: Machine and Organism in Descartes* (Ithaca, NY: Cornell University Press, 2001), pp. 41–45. On debates about the heart, see also François Duchesneau, *Les modèles du vivant de Descartes à Leibniz* (Paris: Vrin, 1998), pp. 55–65; Annie Bitbol-Hespériès, "Cartesian Physiology," in Stephen Gaukroger, John Schuster, and John Sutton, eds., *Descartes' Natural Philosophy* (London: Routledge, 2000), pp. 349–82; Bitbol-Hespériès, "De Vésale à Descartes: Le coeur, la vie," *Histoire des sciences médicales* 48.4 (2014), pp. 513–22. On physiology more generally, see Thomas Steele Hall, *Ideas of Life and Matter: Studies in the History of General Physiology, 600 B.C.–1900 A.D.*, 2 vols. (Chicago: University of Chicago Press, 1969), vol. 1; and Riskin, *Restless Clock*, ch. 2. On Harvey, Descartes, and the circulation of blood, see Thomas Wright, *William Harvey: A Life in Circulation* (Oxford: Oxford University Press, 2013).

30. René Descartes, "Description of the Human Body," in *The Philosophical Writings of Descartes*, trans. John Cottingham, Robert Stoothoff, and Dugald Murdoch. 3 vols. (New York: Cambridge University Press, 1985), vol. 1, p. 316.

31. See Duchesneau, *Les modèles du vivant*, pp. 70–71.

32. See Kaulbach and Meyer, s.v. "Bewegung"; Kaulbach, *Der philosophische Begriff der Bewegung*, p. 26.

33. Kaulbach and Meyer, s.v. "Bewegung," p. 873.

34. In differential calculus, a mathematical curve is divided into a large number of very small straight lines from which the curve's tangent can be derived at any point. In the coordinate system, the small lines are represented as "infinitesimals," and the relationship between them and the continuum is shown visually in the continuous line of the curve and its tangent at different points. On the physics and mathematics of motion in general, see Julian B. Barbour, *Absolute or Relative Motion? Volume 1, The Discovery of Dynamics. A Study from a Machian Point of View of the Discovery and the Structure of Dynamical Theories* (Cambridge: Cambridge University Press, 1989); Michel Blay, *La science du mouvement: De Galilée à Lagrange* (Paris: Belin, 2002); Carla Rita Palmerino and

J. J. M. H. Thijssen, eds., *The Reception of the Galilean Science of Motion in Seventeenth-Century Europe* (Dordrecht: Kluwer, 2004). On the development of mathematics, see Michael S. Mahoney, "Infinitesimals and Transcendent Relations: The Mathematics of Motion in the Late Seventeenth Century," in David C. Lindberg and Robert S. Westman, eds., *Reappraisals of the Scientific Revolution* (Cambridge: Cambridge University Press, 1990), pp. 461–91. On the distinction between mathematicization and Galileo's and Descartes's geometricization, see Michel Blay, *Reasoning with the Infinite: From the Closed World to the Mathematical Universe*, trans. M. B. DeBevoise (Chicago: University of Chicago Press, 1998).

35. Duchesneau, *Les modèles du vivant*, p. 72; see also Guido Giglioni, "Automata Compared: Boyle, Leibniz and the Debate on the Notion of Life and Mind," *British Journal for the History of Philosophy* 3.2 (1995), pp. 249–78; Margaret J. Osler, "Eternal Truths and the Laws of Nature: The Theological Foundations of Descartes' Philosophy of Nature," *Journal of the History of Ideas* 46.3 (1985), pp. 349–62; Des Chene, *Spirits and Clocks*, pp. 17, 25, 68, 72.

36. Not so for Descartes, who showed little interest in actual animals. On the few dissections he performed and attended, see Anita Guerrini, *The Courtiers' Anatomists: Animals and Humans in Louis XIV's Paris* (Chicago: University of Chicago Press, 2015), p. 39.

37. Frampton, *Embodiments of Will*, p. 3.

38. My brief comments here follow the detailed study, based on his own dissections, by Frampton, *Embodiments of Will*, quotations p. 82. On the heart, see p. 31; on the heart and *pneuma*, pp. 81–82; on the *pneuma* more broadly, pp. 68–78.

39. Ibid., pp. 86 and 118.

40. Ibid., pp. 121 and 91; on flesh, see also p. 3.

41. Ibid., pp. 56, 93, 121.

42. Ibid., p. 120. On Herophilus in more detail, see Heinrich von Staden, *Herophilus: The Art of Medicine in Early Alexandria. Edition, Translation and Essays* (Cambridge: Cambridge University Press, 1989); on nerves, esp. pp. 159–61.

43. Phillip de Lacy's translation, quoted by Frampton, *Embodiments of Will*, p. 121; see also Charles Mayo Goss, "On Movement of Muscles by Galen of Pergamon," *American Journal of Anatomy* 123.1 (1968), pp. 1–25.

44. Shigehisa Kuriyama, "'No Pain, No Gain' and the History of Presence," *Representations* 146.1 (2019), p. 96.

45. Galen wrote: "The source of sensation and volition belongs to the controlling

part of the soul" (*Doctrines of Hippocrates and Plato* 8.1, 480), in Frampton, *Embodiments of Will*, p. 159.

46. Ibid., p. 152, quoting Galen, *De nervorum dissectione* 1.2, 831.1–9.

47. Kuriyama, "'No Pain, No Gain,'" p. 97.

48. Julian Jaynes, "The Problem of Animate Motion in the Seventeenth Century," *Journal of the History of Ideas* 31.2 (1970), pp. 219–34; Ugo Baldini, "Animal Motion Before Borelli 1600–1680," in Domenico Bertolini Meli, ed., *Marcello Malpighi: Anatomist and Physician* (Florence: Olschki, 1997), pp. 193–246.

49. Harvey's notes were first published in 1959 in Whitteridge's bilingual edition. William Harvey, *De motu locali animalium, 1627*, ed. and trans. Gweneth Whitteridge (Cambridge: Cambridge University Press, 1959), pp. 45 and 41. The first part of the notes on animate motion consists mainly of a collection of references and commentaries on Aristotle. Of the two sections of Harvey's notebook, the one presented first in Whitteridge's edition is the chronologically older. See Whitteridge, "Introduction," ibid., pp. 2–3.

50. Harvey, *De motu locali animalium*, p. 17.

51. Ibid., p. 15.

52. Giovanni Alfonso Borelli, *On the Movement of Animals*, trans. Paul Maquet (Berlin: Springer, 1989), pp. 7 and 8.

53. My comments on Fabricius follow Distelzweig and his translation, Peter Distelzweig, "Fabricius's Galeno-Aristotelian Teleomechanics of Muscle," in Ohad Nachtomy and Justin E. H. Smith, eds., *The Life Sciences in Early Modern Philosophy* (Oxford: Oxford University Press, 2014), quotation p. 70. Written in 1618, Fabricius's *De motu locali animalium* was part of a larger publication, now lost, the *Totius animalis fabricae theatrum*, and was published in 1625 together with *De musculis* and *De occium articulis*.

54. Harvey, *De motu locali animalium*, p. 67. The three did not agree on whether Aristotle recognized the muscles: Harvey thought he had; Fabricius and Borelli did not. Harvey, *De motu locali animalium*, p. 65; Borelli, *On the Movement of Animals*, p. 15.

55. Distelzweig, "Fabricius's Galeno-Aristotelian Teleomechanics," pp. 67 and 75.

56. The University of Padua, where Fabricius was professor of anatomy from his youth until his old age, 1565–1613, was a strongly Aristotelian environment. See Andrew Cunningham, *The Anatomical Renaissance: The Resurrection of the Anatomical Projects of the Ancients* (Brookfield, VT: Scolar Press, 1997). At Padua, Fabricius was also the physician and colleague of Galileo Galilei, who in 1609 built his own copy of the telescope invented by the Dutch lens grinder Hans Lippershey a year earlier. The results of Galileo's

stargazing were published in his famous *Sidereus Nuncius* (1610), but his planned work on animal motion was never realized. On mutual influences between Galileo, Fabricius, Harvey, and Borelli, see Arturo Castiglioni, "Galileo Galilei and His Influence on the Evolution of Medical Thought," *Bulletin of the History of Medicine* 12.2 (1942), pp. 226–41.

57. This analysis of Fabricius's and Harvey's treatment of muscles is set out in Distelzweig, "Fabricius's Galeno-Aristotelian Teleomechanics," esp. pp. 70–79, and in more detail in Peter M. Distelzweig, "Descartes's Teleomechanics in Medical Context: Approaches to Integrating Mechanics and Teleology in Hieronymus Fabricius ab Aquapendente, William Harvey, and René Descartes," Ph.D. diss., University of Pittsburgh, 2013.

58. Fabricius identifies a specific "nervaceum corpus" as responsible for contraction. See Distelzweig, "Fabricius's Galeno-Aristotelian Teleomechanics," p. 74. Harvey also follows Galen in describing "tonic motion," which is neither contraction or relaxation, but the tension between them. Distelzweig, "Descartes's Teleomechanics," p. 138 and note 192.

59. Borelli, *On the Movement of Animals*, p. 150 (on ice), pp. 197–202 (underwater).

60. See the plates at the end of Borelli, *On the Movement of Animals*, for example, plate 1, fig. 6, and plate 2, fig. 9. On Borelli's mechanism, see Dennis Des Chene, "Mechanisms of Life in the Seventeenth Century: Borelli, Perrault, Régis," *Studies in History and Philosophy of Biology and Biomedical Sciences* 36.2 (2005), pp. 245–60; Sophie Roux, "Quelles machines pour quels animaux? Jacques Rohault, Claude Perrault, Giovanni Alfonso Borelli," in Aurélia Gaillard et al., eds., *L'automate: Machine, métaphore, modèle, merveille* (Bordeaux: Presses universitaires de Bordeaux, 2013), pp. 69–113.

61. Borelli, *On the Movement of Animals*, p. 15, citing Lucretius.

62. Ibid., p. 16. Borelli also experimentally determined the center of gravity in the human body. Ibid., pp. 128–30. This aim was revisited only in the late nineteenth century, by Wilhelm Braune and Otto Fischer.

63. Ibid., p. 2.

64. Domenico Bertoloni Meli, *Mechanism, Experiment, Disease: Marcello Malpighi and Seventeenth-Century Anatomy* (Baltimore: Johns Hopkins University Press, 2011), pp. 280–89.

65. Borelli, *On the Movement of Animals*, pp. 225 and 281–82.

66. Ibid., pp. 282. Borelli interprets fermentation not chemically, but mechanically, as mixing—in other words, as the moving, combining, and separating of particles—though a clear-cut distinction between iatromechanist and iatrochemist positions is difficult to

draw for the late seventeenth century. In 1664 and 1667, the Danish anatomist Niels Stensen had already shown that muscle contraction actually resulted not from swelling, but from the shortening of fibers, a rival model that Borelli rejected. See Troels Kardel, "Niels Stensen's Geometrical Theory of Muscle Contraction (1667): A Reappraisal," *Journal of Biomechanics* 23.10 (1990), pp. 953–65, and Kardel, "Nicolaus Steno's *New Myology* (1667): Rather than Muscle, the Motor Fibre should be Called Animal's Organ of Movement," *Nuncius* 23.1 (2008), pp. 37–64.

67. Borelli, *On the Movement of Animals*, p. 282.

68. Ibid., pp. 282 and 283.

69. Ibid., p. 284.

70. Ibid., p. 285.

71. Ibid., p. 286.

72. Jan Swammerdam, *The Book of Nature; or the History of Insects* (London: C. G. Seyffert, 1758), part 2, p. 128.

73. Ibid., part 2, pp. 123 and 129.

74. For more detail on Haller's physiology, his precursors, and subsequent theories of sensibility and irritability, see Jörg Jantzen, "Physiologische Theorien," in Francesco Moiso, Manfred Durner, and Jörg Jantzen, eds., *Wissenschaftshistorischer Bericht zu Schellings naturphilosophischen Schriften* (Stuttgart: Frommann-Holzboog, 1994), pp. 375–668; Hubert Steinke, *Irritating Experiments: Haller's Concept and the European Controversy on Irritability and Sensibility, 1750–1790* (Amsterdam: Rodopi, 2005); François Duchesneau, *La physiologie des Lumières: Empirisme, modèles et théories* (The Hague: Nijhoff, 1982), ch. 5. On Haller as a polymath, see Richard Toellner, *Albrecht von Haller: Über die Einheit im Denken des letzten Universalgelehrten* (Wiesbaden: Steiner, 1971).

75. Albrecht von Haller, *A Dissertation on the Sensible and Irritable Parts of Animals* (1752), trans. with an introduction by Owsei Temkin (Baltimore: Johns Hopkins Press, 1936), p. 46.

76. Steinke, *Irritating Experiments*, p. 107.

77. See Hall, *Ideas of Life and Matter*, vol. 1, ch. 25 on Stahl and vol. 2, ch. 33–35 on the eighteenth century in general; also Theodore M. Brown, "From Mechanism to Vitalism in Eighteenth-Century English Physiology," *Journal of the History of Biology* 7.2 (1974), pp. 179–216; E. Benton, "Vitalism in Nineteenth-Century Scientific Thought: A Typology and Reassessment," *Studies in History and Philosophy of Science Part A* 5.1 (1974), pp. 17–48; Guido Cimino and François Duchesneau, eds., *Vitalisms from Haller to the Cell Theory*

(Florence: Olschki, 1997); François Duchesneau, "Territoires et frontières du vitalisme 1750–1850," in Guido Cimino and François Duchesneau, eds., *Vitalisms from Haller to the Cell Theory* (Florence: Olschki, 1997), pp. 297–349; Joan Steigerwald, *Experimenting at the Boundaries of Life: Organic Vitality in Germany around 1800* (Pittsburgh: University of Pittsburgh Press, 2019).

CHAPTER TWO: VISUAL EXPERIMENTATION

1. Étienne-Jules Marey, *Du mouvement dans les fonctions de la vie: Leçons faites au Collège de France* (Paris: Germer Baillière, 1868), p. 8.

2. Ibid., pp. 11–12.

3. Carl Friedrich Kielmeyer, *Ideen zu einer allgemeinen Geschichte und Theorie der Entwicklungserscheinungen der Organisationen* (1793–94), quoted in Janina Wellmann, *The Form of Becoming: Embryology and the Epistemology of Rhythm, 1760–1830*, trans. Kate Sturge (New York: Zone Books, 2017), p. 141.

4. See Andrew Cunningham, "The Pen and the Sword: Recovering the Disciplinary Identity of Physiology and Anatomy before 1800, II: Old Anatomy — The Sword," *Studies in History and Philosophy of Biological and Biomedical Sciences* 34.1 (2003), p. 58. On the related concept of "organism" in the wider European linguistic area, see Tobias Cheung, "What is an 'Organism'? On the Occurrence of a New Term and Its Conceptual Transformations 1680–1850," *History and Philosophy of the Life Sciences* 32.2–3 (2010), pp. 155–94. From an epistemological perspective, see Cécilia Bognon-Küss, ed., "Organic, Organization, Organism: Essays in the History and Philosophy of Biology and Chemistry," Topical Collection in *History and Philosophy of the Life Sciences* 38.4 (2016), and Charles T. Wolfe, "Do Organisms Have an Ontological Status?," *History and Philosophy of the Life Sciences* 32.2–3 (2010), pp. 195–231. On the emergence of biology, see Lynn K. Nyhart, *Biology Takes Form: Animal Morphology and the German Universities, 1800–1900* (Chicago: University of Chicago Press, 1995); Jacques Roger, *The Life Sciences in Eighteenth-Century French Thought*, ed. Keith R. Benson, trans. Robert Ellrich (Stanford, CA: Stanford University Press, 1997); John H. Zammito, *The Gestation of German Biology: Philosophy and Physiology from Stahl to Schelling* (Chicago: University of Chicago Press, 2017).

5. See Wellmann, *Form of Becoming*, esp. ch. 6.

6. Marey, *Du mouvement dans les fonctions de la vie*, p. 33.

7. Motion is "one of life's most mysterious functions." Étienne-Jules Marey, *La méthode graphique dans les sciences expérimentales et principalement en physiologie et en*

médecine, 2nd, rev. ed. (Paris: G. Masson, 1885), p. 108. See also Marey, *Du mouvement dans les fonctions de la vie*, p. 203: "the function that, more than any other, characterizes the living being: motricity. When one examines the living organism, one sees that, everywhere, life conveys itself to us by a movement that is more or less perceptible, but always essential to the function that it accompanies."

8. Marey, *Du mouvement dans les fonctions de la vie*, p. 33.

9. Ibid., p. 34.

10. The graphic method is "the only natural method of expressing such events." Étienne-Jules Marey, *Movement*, trans. Eric Pritchard (New York: Appleton, 1895), p. 2.

11. Marey, *La méthode graphique*, p. 124.

12. This was a version of Carl Ludwig's kymograph, invented in 1847. Ibid., p. 112. On numerous other physiological devices, see ibid., pp. 112–16.

13. Ibid., p. iii. See Joel Snyder, "Visualization and Visibility," in Caroline A. Jones and Peter Galison, eds., *Picturing Science, Producing Art* (New York: Routledge, 1998), p. 381; Marta Braun, *Picturing Time: The Work of Etienne-Jules Marey (1830–1904)* (Chicago: University of Chicago Press, 1992), p. x.

14. Lorraine Daston and Peter Galison, *Objectivity* (New York: Zone Books, 2007). From the mid-nineteenth century on, the rise of mechanical objectivity went hand in hand with a scientific culture that attributed a special superiority and even morality to the "machine as a neutral and transparent operator." Peter Galison, "Judgement against Objectivity," in Jones and Galison, eds., *Picturing Science, Producing Art*, p. 332.

15. On the issue of delayed reaction, for example, see Jimena Canales, *A Tenth of a Second: A History* (Chicago: University of Chicago Press, 2009). The literature on experimental physiology and its apparatuses is extensive. For example, Hans-Jörg Rheinberger and Michael Hagner, eds., *Die Experimentalisierung des Lebens: Experimentalsysteme in den biologischen Wissenschaften 1850–1950* (Berlin: Akademie, 1993); Alexandre Métraux and Andreas Mayer, eds., *Kunstmaschinen: Spielräume des Sehens zwischen Wissenschaft und Ästhetik* (Frankfurt am Main: Fischer, 2005). On the towering figure of Hermann von Helmholtz, see David Cahan, *Helmholtz: A Life in Science* (Chicago: University of Chicago Press, 2018); Henning Schmidgen, "Lebensräder, Spektatorien, Zuckungstelegraphen: Zur Archäologie des physiologischen Blicks," in Helmar Schramm, ed., *Bühnen des Wissens: Interferenzen zwischen Wissenschaft und Kunst* (Berlin: Dahlem University Press, 2003), pp. 268–99.

16. Marey, *Du mouvement dans les fonctions de la vie*, p. 25.

17. Marey, *La méthode graphique*, p. 124.

18. Ibid., p. 108; see also pp. vii and 111. Unlike the themes of experiment, chrono-photography, and the graphic method, there has been little scholarly interest either in Marey's physiology as part of the domain inaccessible to the senses or in its consequences for physiological thinking. Exceptions are Josh Ellenbogen, "Camera and Mind," *Representations* 101.1 (2008), pp. 86–115; Ellenbogen, "Educated Eyes and Impressed Images," *Art History* 33.3 (2010) pp. 490–511; Erin Manning, "Grace Taking Form: Marey's Movement Machines," *Parallax* 14.1 (2008), pp. 82–91; Snyder, "Visualization and Visibility"; Braun, *Picturing Time*, pp. xvii–xx and 195; and François Dagognet, *Étienne-Jules Marey: A Passion for the Trace*, trans. Robert Galeta (New York: Zone Books, 1992).

19. See Jonathan Crary, *Suspensions of Perception: Attention, Spectacle, and Modern Culture* (Cambridge, MA: MIT Press, 1999); Crary, *Techniques of the Observer: On Vision and Modernity in the Nineteenth Century* (Cambridge, MA: MIT Press, 1991).

20. See Anson Rabinbach, *The Human Motor: Energy, Fatigue, and the Origins of Modernity* (Berkeley: University of California Press, 1992); Stephen Kern, *The Culture of Time and Space, 1880–1918* (Cambridge, MA: Harvard University Press, 1983). On science, see Nick Hopwood, Simon Schaffer, and James Secord, "Seriality and Scientific Objects in the Nineteenth Century," *History of Science* 48.3–4 (2010), pp. 251–85, and Alex Csiszar, *The Scientific Journal: Authorship and the Politics of Knowledge in the Nineteenth Century* (Chicago: University of Chicago Press, 2018).

21. Ellenbogen, "Camera and Mind," p. 87.

22. Phillip Prodger, "The Romance and Reality of the Horse in Motion," in Joyce Delimata, ed., *Actes du Colloque Marey/Muybridge, pionniers du cinéma: Rencontre Beaune/Stanford* (Beaune: Conseil régional de Bourgogne, 1995), p. 49.

23. On the history of cinematography, the development of optical toys, processes, and projection apparatuses, and the origins of cinema between science and spectacle, see Olaf Breidbach, Kerrin Klinger, and André Karliczek, eds., *Natur im Kasten: Lichtbild, Schattenriss, Umzeichnung und Naturselbstdruck um 1800* (Jena: Ernst-Haeckel-Haus, 2010); Lisa Cartwright, *Screening the Body: Tracing Medicine's Visual Culture* (Minneapolis: University of Minnesota Press, 1995); Scott Curtis, *The Shape of Spectatorship: Art, Science, and Early Cinema in Germany* (New York: Columbia University Press, 2015); Mary Ann Doane, *The Emergence of Cinematic Time: Modernity, Contingency, and the Archive* (Cambridge, MA: Harvard University Press, 2002); Michel Frizot, ed., *A New History of Photography* (Cologne: Könemann, 1998); Oliver Gaycken, *Devices of Curiosity: Early Cinema*

and Popular Science (Oxford: Oxford University Press, 2015); Tom Gunning, "The Cinema of Attractions: Early Film, Its Spectator, and the Avant-Garde," in Thomas Elsaesser, ed., *Early Cinema: Space, Frame, Narrative* (London: BFI Publishing, 1990), pp. 56–62; Paul Liesegang, *Wissenschaftliche Kinematographie: Einschließlich der Reihenphotographie* (Düsseldorf: Liesegang, 1920); Annemone Ligensa and Klaus Kreimeier, eds., *Film 1900: Technology, Perception, Culture* (New Barnet, UK: John Libbey, 2009); Virgilio Tosi, *Cinema before Cinema: The Origins of Scientific Cinematography*, trans. Sergio Angelini (London: British Universities Film & Video Council, 2005); Janina Wellmann, ed., "Cinematography, Seriality, and the Sciences," special issue, *Science in Context* 24.3 (2011).

24. See Prodger, "Romance and Reality." On the iconography of the horse in motion, see Daniel Roche, "Equestrian Culture in France from the Sixteenth to the Nineteenth Century," *Past and Present* 199.1 (2008), pp. 113–45; for the nineteenth century, Andreas Mayer, "The Physiological Circus: Knowing, Representing, and Training Horses in Motion in Nineteenth-Century France," *Representations* 111.1 (2010), pp. 88–120. On art history, see Ernst H. Gombrich, "Moment and Movement in Art," in Gombrich, *The Image and the Eye: Further Studies in the Psychology of Pictorial Representation* (London: Phaidon, 1982). On the topos of the living image in the history of art, see Frank Fehrenbach, "Kohäsion und Transgression: Zur Dialektik des lebendigen Bildes," in Ulrich Pfisterer and Anja Zimmermann, eds., *Transgressionen/Animationen: Das Kunstwerk als Lebewesen* (Berlin: Akademie, 2005), pp. 1–40; Fehrenbach, "'Eine Zartheit am Rande unseres Sehvermögens': Bildwissenschaft und Lebendigkeit," *Kunst und Wissenschaft: Tendenzen, Probleme* 38.3 (2010), pp. 33–44; Fehrenbach, "Quasi animata forma: 'Living Art' in the Early Modern Period," in Marc Wellmann, ed., *BIOS: Konzepte des Lebens in der zeitgenössischen Skulptur* (Berlin: Wienand, 2012), pp. 30–39.

25. Motion photographs quickly became part of the period's visual repertoire. Ottomar Anschütz (1846–1907); Albert Londe (1858–1917); Marey's first assistant, Georges Demeny (1850–1917), and his last assistant, Lucien Bull (1876–1972); the Hungarian artist Bertalan Székely (1835–1919); physicist Ernst Mach (1838–1916); sports scientist Ernst Kohlrausch (1850–1923)—all of these and many more used the technology to explore the world afresh, whether animals in nature, emperors on maneuver, or bullets in flight.

26. Prodger, "Romance and Reality," p. 55.

27. On Muybridge more generally, see Phillip Prodger, *Time Stands Still: Muybridge and the Instantaneous Photography Movement* (Oxford: Oxford University Press, 2003). At the University of Pennsylvania in 1883, Muybridge began his studies for the album *Animal*

Locomotion (1887), which featured 781 plates of motion series taken using a system of multiple cameras. On this technique, see Eadweard Muybridge, *Descriptive Zoopraxography, or, The Science of Animal Locomotion Made Popular* (Philadelphia: University of Pennsylvania, 1893). Many parts of the series used montage techniques, including numbering and reordering, cropping, enlarging, and substituting. See Hans Christian Adam, ed., *Eadweard Muybridge: The Human and Animal Locomotion Photographs* (Cologne: Taschen, 2010); Philip Brookman, ed., *Eadweard Muybridge* (London: Tate, 2010); and Braun, *Picturing Time*, pp. 228–54. Braun characterizes Muybridge's photography as a "narrative tableau," a "treasure trove of figurative imagery, a reiteration of contemporary pictorial practice, and a compendium of social history and erotic fantasy." Ibid., pp. 246 and 249. On the artistic context of motion studies in the United States, see Nancy Mowll Mathews and Charles Musser, eds., *Moving Pictures: American Art and Early Film, 1880–1910* (Manchester, VT: Hudson Hills Press, 2005). On Muybridge, projection devices, and the beginnings of cinema, see Marey, *Movement*, ch. 18; Tosi, *Cinema before Cinema*; Thierry Lefebvre, Jacques Malthête, and Laurent Mannoni, eds., *Sur les pas de Marey: Science(s) et cinéma* (Paris: L'Harmattan, 2004); and Wellmann, "Cinematography, Seriality, and the Sciences." Instructive cultural histories of time and space in the late nineteenth century are Kern, *Culture of Time and Space*, and Canales, *A Tenth of a Second*.

28. See Marey, *Movement*, pp. 103–25. On Janssen's research, also Jimena Canales, "Photogenic Venus: The 'Cinematographic Turn' and Its Alternatives in Nineteenth-Century France," *Isis* 93.4 (2002), pp. 585–613.

29. Marey, *La méthode graphique*, supplement "Développement de la méthode graphique par l'emploi de la photographie," p. 3.

30. Marey, *Movement*, p. 19. Joel Snyder discusses Marey's inconsistencies in describing photography: struggling "to establish the boundaries separating visualization and visibility," Marey "never gave up the conception of photography as representing what we see, but neither did he give up the notion of chronophotography as a form of revelation of the imperceptible." Snyder, "Visualization and Visibility," pp. 395 and 393.

31. Marey, *Movement*, p. 18.

32. Ibid., p. 34; also Marey, *La méthode graphique*, supplement, p. 2.

33. Marey, *La méthode graphique*, supplement, p. 2.

34. See Marey, *Movement*, pp. 2, 8, 11, 22, 63, 64.

35. Ibid., p. 2.

36. Ibid., quotation p. 20. Whereas Descartes's mathematical curves were a "laborious

construction," comments Marey, his own graphic method produces lines automatically, traced by styli and rotating cylinders. Ibid., p. 34; on Descartes, see also Marey, *La méthode graphique*, pp. iv and 11–12.

37. Marey, *Movement*, p. 54.

38. Ibid., p. 34.

39. Ibid., pp. 33–34; see also p. 57.

40. Marey, *La méthode graphique*, supplement, pp. 22–37.

41. See ibid., supplement, pp. 32–37, quotation p. 33; Marey, *Movement*, pp. 57–61.

42. See Étienne-Jules Marey, *La machine animale: Locomotion terrestre et aérienne* (Paris: Germer Baillière, 1873), book 1; Marey, *Movement*, p. 34.

43. See Braun, *Picturing Time*, ch. 8; Rabinbach, *Human Motor*; Hans-Christian von Herrmann, "Motion Records: Die Bewegung im Zeitalter ihrer technischen Informierbarkeit," in Claudia Jeschke and Hans-Peter Bayerdörfer, eds., *Bewegung im Blick: Beiträge zu einer theaterwissenschaftlichen Bewegungsforschung* (Berlin: Vorwerk 8, 2000), pp. 100–12.

44. Wilhelm Weber and Eduard Weber, *Mechanics of the Human Walking Apparatus* (1836), trans. P. Maquet and R. Furlong (Berlin: Springer, 1992). The Weber brothers' work already reduces the figure in motion to a silhouette (which, unlike in Marey's work, is hand drawn). The Webers also used a zoetrope to animate movements. However, they failed to win support for their approach of ignoring the musculature and regarding the leg as a freely swinging pendulum. See Andreas Mayer, *The Science of Walking: Investigations into Locomotion in the Long Nineteenth Century*, trans. Tilman Skowroneck and Robin Blanton (Chicago: University of Chicago Press, 2020); Wilhelm Braune and Otto Fischer, "Der Gang des Menschen: I. Theil: Versuche am unbelasteten und belasteten Menschen," *Abhandlungen der Königlich Sächsischen Gesellschaft der Wissenschaften: Mathematisch-physische Classe* 21 (1895), pp. 151–322. A history of locomotion research can be found in the dissertation of Marey's student Gaston Carlet, *Essai expérimental sur la locomotion humaine: Étude de la marche* (Paris: Martinet, 1872), ch. 1.

45. See Mayer, *Science of Walking*, for a discussion of the human gait within a specifically French discourse that blended literature and science, culture and laboratory, vitalism and experiment.

46. Marey, *Movement*, p. 311.

47. Ibid., p. 304.

48. Manning, "Grace Taking Form," pp. 82 and 83.

49. Ibid., p. 90.

50. Ibid., p. 83.

51. This moment is James Harris's *punctum temporis* or the "pregnant moment" in Lessing's *Laocoon*, which anticipates the next moment and can therefore never be the climax of an action. Gombrich comments that the idea "is not only an absurdity logically, it is a worse absurdity psychologically." Gombrich, "Moment and Movement," p. 45. Chronophotography, in contrast, represented the nonsimultaneous simultaneously within a single image and thus for the first time offered pictorial resources that went beyond perspective. For the influence of this innovation on art, see Braun, *Picturing Time*, ch. 7.

52. Manning, "Grace Taking Form," p. 85.

53. Crary, *Suspensions of Perception*, quotation p. 13.

54. Jules Janssen and Alphonse Davanne, quoted in Ellenbogen, "Camera and Mind," p. 112 n.3. For Marey, this was the basis of science's mission and especially its devices for the visual education of the eye. Ellenbogen, "Educated Eyes."

CHAPTER THREE: CALCULATING PERCEPTION

1. James E. Cutting, "Gunnar Johansson, Events, and Biological Motion," in Kerri L. Johnson and Maggie Shiffrar, eds., *People Watching: Social, Perceptual, and Neurophysiological Studies of Body Perception* (Oxford: Oxford University Press, 2013), p. 11. The two texts were Gunnar Johansson, "Visual Perception of Biological Motion and a Model for Its Analysis," *Perception & Psychophysics* 14.2 (1973), pp. 201–11, and Johansson, "Spatio-Temporal Differentiation and Integration in Visual Motion Perception," *Psychological Research* 38.4 (1976), pp. 379–93.

2. Johansson's films have been digitized and can be found online. See "2-Dimensional Motion Perception," www.youtube.com/watch?v=1F5ICP9SYLU, www.youtube.com/watch?v=KT89CQ2nRpo. More recent demonstrations of this kind of experiment can be accessed on James Cutting's Cornell website, http://people.psych.cornell.edu/~jec7/biological_motion.htm. I thank James Cutting for these references. On other information obtainable from point-light displays apart from the perception of an articulated shape in motion, see Nikolaus F. Troje and Dorita H. F. Chang, "Shape-Independent Processing of Biological Motion," in Johnson and Shiffrar, eds., *People Watching*, pp. 82–99. Johansson and many others concentrated on human locomotion, and it was only in the 1990s that the first work appeared on people's ability to recognize animals on the basis of their biological motion, as well, for example, George Mather and Sophie West,

"Recognition of Animal Locomotion from Dynamic Point-Light Displays," *Perception* 22.7 (1993), pp. 759–66. Mather and West animated Muybridge's animal motion sequences on the computer, highlighted the animals' joints with bright points, blanked out the rest of the animal, and showed the resulting animations to their test subjects. A related, currently very significant field is computer animation, which asks how movements inscribe information about emotions, identity, and personality.

3. Johansson, "Visual Perception of Biological Motion," p. 201.

4. Cutting, "Gunnar Johansson," p. 15.

5. Gunnar Johansson, *Configurations in Event Perception: An Experimental Study* (Uppsala: Almqvist & Wiksell, 1950), pp. 12–13.

6. Ibid., pp. 11–12.

7. Gunnar Johansson, "Visual Vector Analysis and the Optic Sphere Theory," in Gunnar Jansson, Sten Sture Bergström, and William Epstein, eds., *Perceiving Events and Objects* (Hillsdale, NJ: Lawrence Erlbaum, 1994), p. 270.

8. On Johansson's biography, the scope of his research interests, and his work on various Gestalt theorists, see Sheena Rogers, "Gunnar Johansson: A Practical Theorist: An Interview with William Epstein," in Jansson, Bergström, and Epstein, eds., *Perceiving Events and Objects*, pp. 3–25.

9. Johansson, *Configurations in Event Perception*, p. 18. See also Cutting, "Gunnar Johansson," p. 15.

10. Johansson, "Visual Perception of Biological Motion," p. 205. "Rigidity," a term used in visual perception research, refers to the assumption that in the human perceptual system, objects remain the same objects despite all the transformations (in shape, space, or retinal location due to the movements of the eye) they undergo.

11. Gunnar Johansson, "Rigidity, Stability, and Motion in Perceptual Space: A Discussion of Some Mathematical Principles Operative in the Perception of Relative Motion and of Their Possible Function as Determinants of a Static Perceptual Space," *Acta Psychologica* 14 (1958), p. 360.

12. Johansson, "Visual Perception of Biological Motion," p. 205.

13. Gunnar Johansson, "Vector Analysis in Visual Perception of Rolling Motion," *Psychologische Forschung* 36.4 (1974), p. 312; see also Johansson, "Rigidity, Stability, and Motion."

14. Mark H. Johnson, "Biological Motion: A Perceptual Life Detector?," *Current Biology* 16.10 (2006), pp. R376–R377. Another field of research seeks to discover whether

biological motion is processed in localizable regions of the human brain. For an overview of the abundant literature, see Johnson and Shiffrar, eds., *People Watching*; M. D. Rutherford and Valerie A. Kuhlmeier, eds., *Social Perception: Detection and Interpretation of Animacy, Agency, and Intention* (Cambridge, MA: MIT Press, 2013); and Troje and Chang, "Shape-Independent Processing of Biological Motion." The assumption that biological motion is a distinct perceptual category is contested by Arieta Chouchourelou, Alissa Golden, and Maggie Shiffrar, "What Does 'Biological Motion' Really Mean? Differentiating Visual Percepts of Human, Animal, and Nonbiological Motions," in Johnson and Shiffrar, eds., *People Watching*, pp. 63–81.

15. Nikolaus F. Troje, "What Is Biological Motion? Definition, Stimuli, Paradigms," in Rutherford and Kuhlmeier, eds., *Social Perception*, p. 15.

16. Cutting, "Gunnar Johansson," p. 18.

17. Johansson, "Visual Perception of Biological Motion," p. 210; see also Rogers, "Gunnar Johansson," pp. 17 and 19.

18. Johansson, "Visual Vector Analysis and the Optic Sphere Theory," p. 270 (original emphasis).

19. Johansson, "Rigidity, Stability, and Motion," p. 366.

20. "The percepts correspond to the results of such vector analysis along a common time axis; we find that all vectors, which are equal at a given moment, and thus have the same size and direction, combine perceptually into units of motion." Ibid., p. 361.

21. Ibid.

22. Erin Manning, "Grace Taking Form: Marey's Movement Machines," *Parallax* 14.1 (2008), pp. 90 and 83; Johansson, "Rigidity, Stability, and Motion," p. 369.

23. All quotations from Johansson, "Rigidity, Stability, and Motion," p. 369.

24. Johansson, "Visual Vector Analysis and the Optic Sphere Theory," pp. 270–71. Despite this change, Johansson remarks elsewhere that even textbooks fail to show that the visual information is not a matter of "moving images," but the basis of a new mathematical theory. Rogers, "Gunnar Johansson," pp. 19–20.

25. A development that is less relevant for this book, but was greatly influenced by Johansson's work, is the rise of motion capture technologies. Of the rich literature on this subject, I would merely point to Nicolás Salazar Sutil's cultural theory of technological motion, in Sutil, *Motion and Representation: The Language of Human Movement* (Cambridge, MA: MIT Press, 2015).

CHAPTER FOUR: SILICONE MOTION

1. James Gorman, "This Tiny Robot Walks, Crawls, Jumps and Swims. But It Is Not Alive," *New York Times*, January 24, 2018, www.nytimes.com/2018/01/24/science/tiny-robot-medical.html.

2. Wenqi Hu et al., "Small-Scale Soft-Bodied Robot with Multimodal Locomotion," *Nature* 554 (2018), p. 81.

3. Ibid.

4. Ibid., p. 82. See the videos at https://www.nature.com/articles/nature25443?proof=t#Sec4.

5. *Oxford Dictionaries*, s.v. "robot," https://en.oxforddictionaries.com/definition/robot; *Duden Wörterbuch* online, s.v. "Roboter," www.duden.de/rechtschreibung/Roboter; *Larousse* online, s.v. "robot," www.larousse.fr/dictionnaires/francais/robot/69647?q=robot#68894.

6. See Chapter 1, section "Figurations of Artifical Motion."

7. The literature on robotics is enormous. Just as numerous are the various, though increasingly converging approaches in the fields of biomimetic locomotion, soft robotics, nanorobotics, and machine intelligence as represented in specialized publications such as *International Journal of Robotics Research*, *Transactions on Robotics*, and *Soft Robotics*. For overviews, see Jeffrey Aguilar et al., "A Review on Locomotion Robophysics: The Study of Movement at the Intersection of Robotics, Soft Matter, and Dynamical Systems," *Reports on Progress in Physics* 79.11 (2016), 110001; Jinxing Li et al., "Micro/nanorobots for Biomedicine: Delivery, Surgery, Sensing, and Detoxification," *Science Robotics* 2.4 (2017), eaam6431. For a critical view of the term "nanomachine," see Bernadette Bensaude-Vincent and Xavier Guchet, "Nanomachine: One Word for Three Different Paradigms," *Techné* 11.1 (2007), pp. 71–89. On the background, see Bernadette Bensaude-Vincent, "Reconfiguring Nature through Syntheses: From Plastics to Biomimetics," in Bernadette Bensaude-Vincent and William R. Newman, eds., *The Artificial and the Natural: An Evolving Polarity* (Cambridge, MA: MIT Press, 2007), pp. 294–312.

8. Hu et al., "Small-Scale Soft-Bodied Robot," p. 81.

9. Eukaryotic cells have a nucleus, a differentiated internal structure, and organelles. The interior of the cell is the cytoplasm, composed of a more fluid part containing dissolved proteins and other substances (the cytosol) and a more solid part (the cytoskeleton).

10. Over the past twenty years, research on the conversion of chemical energy into

mechanical work and motion through the consumption of ATP (adenosine triphosphate) on a cellular and molecular level has brought together the experimental and theoretical interests of many different disciplines. Soft-matter physics, chemistry, cell biology, molecular engineering, materials science, precision medicine, and synthetic biology all—from different, yet related perspectives—investigate biological motor proteins, synthetic molecular motors, active matter, biohybrid microrobots, and nanomotors. They ask how the natural motors of the organism generate movement and whether that movement may be reproduced synthetically. Of the abundant literature on motor proteins, see the review by Anatoly B. Kolomeisky, "Motor Proteins and Molecular Motors: How to Operate Machines at the Nanoscale," *Journal of Physics: Condensed Matter* 25.46 (2013), 463101; on active matter, the collection "Active Matter," in *Nature*, www.nature.com/collections/ hvczfmjfzl?utm_source=twitter&utm_medium=social&utm_content=boosted&utm_ campaign=NCOM_1_SZ_ActiveMatter-GRC-social; Bernadette Bensaude-Vincent, "From Self-Organization to Self-Assembly: A New Materialism?," *History and Philosophy of the Life Sciences* 38.3 (2016), 1; Evelyn Fox Keller, "Active Matter, Then and Now," *History and Philosophy of the Life Sciences* 38.3 (2016), 11.

11. Swimming motion at low Reynolds numbers is a case that has been studied from the 1950s to the present. See Geoffrey Ingram Taylor, "Analysis of the Swimming of Microscopic Organisms," *Proceedings of the Royal Society of London. Series A. Mathematical and Physical Sciences* 209.1099 (1951), pp. 447–61; E. M. Purcell, "Life at Low Reynolds Number," *American Journal of Physics* 45.3 (1977), pp. 3–11; Alfred Shapere and Frank Wilczek, "Geometry of Self-Propulsion at Low Reynolds Number," *Journal of Fluid Mechanics* 198 (1989), pp. 557–85; Netta Cohen and Jordan H. Boyle, "Swimming at Low Reynolds Number: A Beginners Guide to Undulatory Locomotion," *Contemporary Physics* 51.2 (2010), pp. 103–23; Jake J. Abbott et al., "How Should Microrobots Swim?," *International Journal of Robotics Research* 28.11–12 (2009), pp. 1434–47.

12. Pelin Erkoc et al., "Mobile Microrobots for Active Therapeutic Delivery," *Advanced Therapeutics* 2.1 (2019), 1800064, p. 2.

13. Ibid., p. 1. My comments rely on the survey of different approaches, further reading, and future perspectives for this complex and rapidly developing field provided by Yunus Alapan et al., "Microrobotics and Microorganisms: Biohybrid Autonomous Cellular Robots," *Annual Review of Control, Robotics, and Autonomous Systems* 2.1 (2019), pp. 205–30.

14. Like macrophages, neutrophils are part of the innate immune system, whereas

lymphocytes belong to the adaptive immune system, responding to specific pathogens. Much about neutrophils is still unknown or highly contested. See Andrés Hidalgo et al., "The Neutrophil Life Cycle," *Trends in Immunology* 40.7 (2019), pp. 584–97.

15. Ibid., p. 584.

16. For more detail, see Peter Friedl, Stefan Borgmann, and Eva-B. Bröcker, "Amoeboid Leukocyte Crawling through Extracellular Matrix: Lessons from the Dictyostelium Paradigm of Cell Movement," *Journal of Leukocyte Biology* 70.4 (2001), pp. 491–509, and Alapan et al., "Microrobotics and Microorganisms."

17. Erkoc et al., "Mobile Microrobots," p. 14.

18. Klaus Ley et al., "Getting to the Site of Inflammation: The Leukocyte Adhesion Cascade Updated," *Nature Reviews Immunology* 7 (2007), p. 679.

19. See Friedl, Borgmann, and Bröcker, "Amoeboid Leukocyte Crawling," p. 503.

20. Erkoc et al., "Mobile Microrobots," p. 7.

21. See ibid.; Alapan et al., "Microrobotics and Microorganisms," p. 215.

22. Demir Akin et al., "Bacteria-Mediated Delivery of Nanoparticles and Cargo into Cells," *Nature Nanotechnology* 2 (2007), p. 441. The video material can be viewed at https://www.nature.com/articles/nnano.2007.149#Sec15.

23. Erkoc et al., "Mobile Microrobots," p. 5; see also Alapan et al., "Microrobotics and Microorganisms," pp. 220–22.

24. Alapan et al., "Microrobotics and Microorganisms," p. 210. For more detail, see Howard C. Berg, "The Rotary Motor of Bacterial Flagella," *Annual Review of Biochemistry* 72 (2003), pp. 19–54; Rasika M. Harshey, "Bacterial Motility on a Surface: Many Ways to a Common Goal," *Annual Review of Microbiology* 57 (2003), pp. 249–73; T. L. Jahn and E. C. Bovee, "Movement and Locomotion of Microorganisms," *Annual Review of Microbiology* 19 (1965), pp. 21–58.

25. Erkoc et al., "Mobile Microrobots," p. 8; see also Xiaohui Yan et al., "Multifunctional Biohybrid Magnetite Microrobots for Imaging-Guided Therapy," *Science Robotics* 2.12 (2017), eaaq1155.

26. Alapan et al., "Microrobotics and Microorganisms," p. 211.

27. Douglas B. Weibel et al., "Microoxen: Microorganisms to Move Microscale Loads," *Proceedings of the National Academy of Sciences of the United States of America* 102.34 (2005), p. 11963.

28. Haifeng Xu et al., "Sperm-Hybrid Micromotor for Targeted Drug Delivery," *ACS Nano* 12.1 (2018), p. 327.

29. Reza Nosrati et al., "Two-Dimensional Slither Swimming of Sperm within a Micrometre of a Surface," *Nature Communications* 6 (2015), p. 8703.

30. Alapan et al., "Microrobotics and Microorganisms," p. 210.

31. Xu et al., "Sperm-Hybrid Micromotor."

32. The rockets move using "catalytic bubble propulsion," function at body temperature, and are guided by magnetic actuation. Zhiguang Wu et al., "Biodegradable Protein-Based Rockets for Drug Transportation and Light-Triggered Release," *ACS Applied Materials & Interfaces* 7.1 (2015), pp. 250–55; Erkoc et al., "Mobile Microrobots," p. 8; Alexander A. Solovev et al., "Catalytic Microtubular Jet Engines Self-Propelled by Accumulated Gas Bubbles," *Small* 5.14 (2009), pp. 1688–92.

CHAPTER FIVE: PERFORMING ORGANISMS

1. The classic studies include Savile Bradbury, *The Evolution of the Microscope* (Oxford: Pergamon, 1967); Gerard L'Estrange Turner, "Microscopical Communication," ch. 11 of *Essays on the History of the Microscope* (Oxford: Senecio, 1980), who writes: "There was great curiosity concerning all manner of things that were examined whole. It can scarcely be claimed that any research was done" (p. 216). See also C. H. Lüthy, "Atomism, Lynceus, and the Fate of Seventeenth-Century Microscopy," *Early Science and Medicine* 1.1 (1996), pp. 1–27. Amendments to this dismissive view are offered by, especially, Marc J. Ratcliff, *The Quest for the Invisible: Microscopy in the Enlightenment* (Farnham, UK: Ashgate, 2009); Ratcliff, "Wonders, Logic, and Microscopy in the Eighteenth Century: A History of the Rotifer," *Science in Context* 13.1 (2000), pp. 93–119; Ratcliff, "Temporality, Sequential Iconography and Linearity in Figures: The Impact of the Discovery of Division in Infusoria," *History and Philosophy of the Life Sciences* 21.3 (1999), pp. 255–92; James Elkins, "On Visual Desperation and the Bodies of Protozoa," *Representations* 40 (Autumn 1992), pp. 33–55; Barbara Maria Stafford, "Images of Ambiguity: Eighteenth-Century Microscopy and the Neither/Nor," in David Philip Miller and Hanns Peter Reill, eds., *Visions of Empire: Voyages, Botany, and Representations of Nature* (Cambridge: Cambridge University Press, 1996) pp. 230–57; Marian Fournier, *The Fabric of Life: Microscopy in the Seventeenth Century* (Baltimore, MD: Johns Hopkins University Press, 1996).

2. Antoni van Leeuwenhoek, "Observations, communicated to the Publisher by Mr. Antony van Leewenhoeck, in a Dutch Letter of the 9th of Octob. 1676, here English'd: Concerning little Animals by him observed in Rain- Well- Sea- and Snow-water; as also in water wherein Pepper had lain infused," *Philosophical Transactions of the Royal*

Society of London 12.133 (1677), p. 821. It was on September 7, 1674, tucked away at the end of another long letter, that Leeuwenhoek first mentioned having studied lake water in which "there crawled abundance of little animals." Leeuwenhoek, "More Observations from Mr. Leewenhook, in a Letter of Sept. 7. 1674, sent to the Publisher," *Philosophical Transactions of the Royal Society of London* 9.108 (1674), p. 182.

3. Because the published version of the 1676 letter was the one familiar to Leeuwenhoek's contemporaries, here and in most other cases, I quote directly from the *Philosophical Transactions*. Not until 1685 did Leeuwenhoek begin to publish collections of selected letters.

4. Leeuwenhoek, "Observations," pp. 823–27.

5. Ibid., pp. 827 and 828.

6. These are reported in Antoni van Leeuwenhoek, "An Abstract of a Letter from Mr. Anthony Leevvenhoeck at Delft, dated Sep. 17. 1683, containing some Microscopical Observations, about Animals in the scurf of the Teeth, the substance call'd Worms in the Nose, the Cuticula consisting of Scales," *Philosophical Transactions of the Royal Society of London* 14.159 (1684), pp. 568–74, and in the original Dutch version of the letter, Leeuwenhoek-Commissie, ed., *Alle de brieven van Antoni van Leeuwenhoek / The Collected Letters of Antoni van Leeuwenhoek*, 15 vols. (Amsterdam: Swets & Zeitlinger, 1939), vol. 2, pp. 135–37, 151–55.

7. Leeuwenhoek, "Observations," p. 821. The reference to the horse's ears is taken from the full, almost twenty-page version of the letter, translated and published by the protozoologist Clifford Dobell for the first time in 1932. Dobell undertook a microbiological classification of the animalcula based on Leeuwenhoek's description of their form and motion and confirmed that Leeuwenhoek had seen the green algae *Spirogyra*, rotifera (wheel animals), protozoa such as *Vorticella* and Ciliophora, and bacteria. Dobell, *Antony van Leeuwenhoek and His "Little Animals"; Being some Account of the Father of Protozoology and Bacteriology and his Multifarious Discoveries in these Disciplines* (New York: Harcourt, Brace, 1932), p. 118.

8. Leeuwenhoek, "Observations," pp. 821 and 822.

9. Ibid., p. 824.

10. Ibid., pp. 827 and 828.

11. For example, ibid., pp. 830 and 831.

12. Dobell, *Antony van Leeuwenhoek*, pp. 140–41, 142, 143.

13. Ibid., p. 149.

14. Ibid., p. 154.

15. Ibid., p. 144. Robert Hooke, who succeeded Oldenburg as secretary of the Royal Society in 1677, repeated Leeuwenhoek's experiments himself. Only at the third attempt was he able to confirm that there are animals "soe perfectly shaped & indeed with such curious organs of motion." Hooke to Leeuwenhoek, quoted in ibid., p. 183.

16. On the history of microscopy, see Edward G. Ruestow, *The Microscope in the Dutch Republic: The Shaping of Discovery* (Cambridge: Cambridge University Press, 1996); Catherine Wilson, *The Invisible World: Early Modern Philosophy and the Invention of the Microscope* (Princeton, NJ: Princeton University Press, 1995); Fournier, *Fabric of Life*; Brian J. Ford, *Single Lens: The Story of the Simple Microscope* (New York: Harper & Row, 1985). On the popularization of microscopy in England and women's interest in the technique, see Marjorie Hope Nicolson, *Science and Imagination* (Ithaca, NY: Great Seal Books, 1956).

17. On Leeuwenhoek, see Dobell, *Antony van Leeuwenhoek*; Lesley Robertson et al., *Antoni van Leeuwenhoek: Master of the Minuscule* (Leiden: Brill, 2016); Ruestow, *Microscope*, chs. 6 and 7; J. R. Porter, "Antony van Leeuwenhoek: Tercentenary of his Discovery of Bacteria," *Bacteriological Reviews* 40.2 (1976), pp. 260–69. On his single lens microscopes and the production of glass beads, see Brian Bracegirdle, *Beads of Glass: Leeuwenhoek and the Early Microscope* (Leiden: Museum Boerhaave, 1983); Huib J. Zuidervaart and Douglas Anderson, "Antony van Leeuwenhoek's Microscopes and Other Scientific Instruments: New Information from the Delft Archives," *Annals of Science* 73.3 (2016), pp. 257–88. On Leeuwenhoek and dark-field illumination in microscopy, see Robertson, *Antoni van Leeuwenhoek*, ch. 4; Barnett Cohen, "On Leeuwenhoek's Method of Seeing Bacteria," *Journal of Bacteriology* 34.3 (1937), pp. 343–46. On his connections with the guilds, the arts, and especially with Vermeer, see Laura J. Snyder, *Eye of the Beholder: Johannes Vermeer, Antoni van Leeuwenhoek, and the Reinvention of Seeing* (New York: Norton, 2015). On his studies of microorganisms, see Dobell, *Antony van Leeuwenhoek*; Nick Lane, "The Unseen World: Reflections on Leeuwenhoek (1677) 'Concerning Little Animals,'" *Philosophical Transactions of the Royal Society B: Biological Sciences* 370 (2015), 20140344; Robertson et al., *Antoni van Leeuwenhoek*, ch. 5; Porter, "Antony van Leeuwenhoek"; Fournier, *Fabric of Life*, pp. 167–77.

18. Constantijn Huygens to the Royal Society, August 8, 1673, quoted in Dobell, *Antony van Leeuwenhoek*, p. 43.

19. "The technology of virtual witnessing involves the production in a *reader's* mind of such an image of an experimental scene as obviates the necessity for either direct

witness or replication." Steven Shapin and Simon Schaffer, *Leviathan and the Air Pump: Hobbes, Boyle, and the Experimental Life* (Princeton, NJ: Princeton University Press, 1985), p. 60. On virtual witnessing with specific reference to microscopy, see Wilson, *Invisible World*, pp. 98–102.

20. See Dobell, *Antony van Leeuwenhoek*, pp. 167–74; Robertson et al., *Antoni van Leeuwenhoek*, pp. 82–86.

21. Leeuwenhoek, "Observations," p. 821.

22. For example, ibid., pp. 821 and 827. The Dutch originals of Leeuwenhoek's letters (with English translations) are collected in Leeuwenhoek-Commissie, *Alle de brieven*.

23. Leeuwenhoek, "Observations," pp. 827 and 823.

24. The motion and animality of animalcula is mentioned in the standard historical accounts, but taken for granted or at most discussed as a source of delight for the microscopists. See Ruestow, *Microscope*, pp. 178–80; Fournier, *Fabric of Life*, pp. 172–73; Lane, "Unseen World." On scientific preconceptions, see Edward G. Ruestow, "Images and Ideas: Leeuwenhoek's Perception of the Spermatozoa," *Journal of the History of Biology* 16.2 (1983), p. 203.

25. Unpublished passage from Leeuwenhoek's 1676 letter, in Dobell, *Antony van Leeuwenhoek*. p. 141.

26. Unpublished passage from Leeuwenhoek's 1676 letter, ibid., p. 151.

27. From a Royal Society manuscript, quoted in ibid., p. 175.

28. Leeuwenhoek to Robert Hooke, November 12, 1680, ibid., p. 201.

29. Reinier de Graaf, *De mulierum organis generationi inservientibus*, with an introduction by J. A. Dongen (Nieuwkoop: B. de Graaf, 1965). The ovum itself was discovered by Karl Ernst von Baer in 1827. On early theories of generation, see Jacques Roger, *The Life Sciences in Eighteenth-Century French Thought*, ed. Keith R. Benson, trans. Robert Ellrich (Stanford, CA: Stanford University Press, 1997), esp. pp. 236–58 on animalculism; Francis Joseph Cole, *Early Theories of Sexual Generation* (Oxford: Clarendon Press of Oxford University Press, 1930); Joseph Needham, *A History of Embryology* (Cambridge: Cambridge University Press, 1959); Howard B. Adelmann, *Marcello Malpighi and the Evolution of Embryology*, 5 vols. (Ithaca, NY: Cornell University Press, 1966); John Farley, *The Spontaneous Generation Controversy from Descartes to Oparin* (Baltimore, MD: Johns Hopkins University Press, 1979); Farley, *Gametes and Spores: Ideas about Sexual Reproduction, 1750–1914* (Baltimore, MD: Johns Hopkins University Press, 1982); Edmund O. von Lippmann, *Urzeugung und Lebenskraft: Zur Geschichte dieser Probleme von den ältesten Zeiten*

an bis zu den Anfängen des 20. Jahrhunderts (Berlin: J. Springer, 1933); Wilson, *Invisible World*, esp. ch. 4.

30. Later, this view led Leeuwenhoek to what he thought was an observation of future living beings in the sperm. See Ruestow, "Images and Ideas." On Leeuwenhoek's experiments, see Carlo Castellani, "Spermatozoan Biology from Leeuwenhoek to Spallanzani," *Journal of the History of Biology* 6.1 (1973), pp. 37–68.

31. Antoni van Leeuwenhoek, "An Abstract of a Letter of Mr. Leeuwenhoeck Fellow of the R. Society, dated March 30th. 1685, to the R. S. Concerning Generation by an Insect," *Philosophical Transactions of the Royal Society of London* 15.174 (1685), p. 1120.

32. Ibid., pp. 1120–21, 1124, 1121.

33. Ibid., p. 1120.

34. Ibid., p.1123.

35. Ibid., p. 1126.

36. From the wealth of literature on the visual culture of natural history and early modern knowledge, some examples will indicate the breadth of research: Helen Anne Curry et al., eds., *Worlds of Natural History* (Cambridge: Cambridge University Press, 2018); Pamela H. Smith and Paula Findlen, eds., *Merchants and Marvels: Commerce, Science, and Art in Early Modern Europe* (London: Routledge, 2002); Pamela H. Smith, *The Body of the Artisan: Art and Experience in the Scientific Revolution* (Chicago: University of Chicago Press. 2004); Ann Bermingham, *Learning to Draw: Studies in the Cultural History of a Polite and Useful Art* (New Haven, CT: Yale University Press, 2000); Horst Bredekamp, *The Lure of Antiquity and the Cult of the Machine: The Kunstkammer and the Evolution of Science, Art, and Technology*, trans. Allison Brown (Princeton, NJ: M. Wiener, 1995); Horst Bredekamp, Jochen Brüning, and Cornelia Weber, eds., *Theater der Natur und Kunst = Theatrum naturae et artis: Wunderkammern des Wissens* (Berlin: Henschel, 2000); Caroline A. Jones and Peter Galison, eds., *Picturing Science, Producing Art* (New York: Routledge, 1998); Daniela Bleichmar, *Visible Empire: Botanical Expeditions and Visual Culture in the Hispanic Enlightenment* (Chicago: University of Chicago Press, 2012); Svetlana Alpers, *The Art of Describing: Dutch Art in the Seventeenth Century* (Chicago: University of Chicago Press, 1983). With special attention to the visual world of microscopy, see David Freedberg, "Iconography between the History of Art and the History of Science: Art, Science, and the Case of the Urban Bee," in Caroline A. Jones and Peter Galison, eds., *Picturing Science, Producing Art* (New York: Routledge, 1998), pp. 272–96; Freedberg, *The Eye of the Lynx: Galileo, His Friends, and the Beginnings of Modern Natural History* (Chicago:

University of Chicago Press, 2002); Meghan C. Doherty, "Discovering the 'True Form': Hooke's *Micrographia* and the Visual Vocabulary of Engraved Portraits," *Notes & Records of the Royal Society* 66.3 (2012), pp. 211–34.

37. Leewenhoeck, "An Abstract of a Letter containing some Microscopical Observations," pp. 568–69. On the representation of sperm in the seventeenth century, see A. W. Meyer, "The Discovery and Earliest Representations of Spermatozoa," *Bulletin of the Institute of the History of Medicine* 6.2 (1938), pp. 89–110.

38. Antoni van Leeuwenhoek, "IV. Part of a Letter from Mr Antony van Leeuwenhoek, F. R. S. concerning green Weeds growing in Water, and some Animalcula found about them," *Philosophical Transactions of the Royal Society of London* 23.283 (1703), p. 1305.

39. Ibid., p. 1306.

40. Antoni van Leeuwenhoek, "II. An Extract of a Letter from Mr. Leewenhoek, Dated the 10th of July, An. 1695. Containing Microscopical Observations on Eels, Mites, the Seeds of Figs, Strawberries, &c.," *Philosophical Transactions of the Royal Society of London* 19.221 (1695), p. 271.

41. From an unpublished letter of September 14, 1694, in Leeuwenhoek-Commissie, *Alle de brieven*, vol. 10, p. 149.

42. Leeuwenhoek, "II. An Extract of a Letter," p. 272.

43. The drawing, commissioned by Leeuwenhoek, is reproduced in Leeuwenhoek-Commissie, *Alle de brieven*, vol. 12, plate 4. For a detailed analysis of the plate, see Karin Leonhard, "Blut sehen," in Inge Hinterwaldner and Markus Buschhaus, eds., *The Picture's Image: Wissenschaftliche Visualisierung als Komposit* (Munich: Fink, 2006), pp. 104–28. On Leeuwenhoek's draftsmen, whom he never names, see also Dobell, *Antony van Leeuwenhoek*, pp. 342–45. Leeuwenhoek did not regard the blood's corpuscles as alive, despite all their movements, because their motion was not self-propelled, but induced mechanically from the outside. On the relationship between corpuscular theories of matter and seventeenth-century microscopy, see Wilson, *Invisible World*, and Lüthy, "Atomism."

44. Leonhard, "Blut sehen," p. 122.

45. As is so often the case, the disciplinary boundaries between art history, the history of science, the history of technology, and media history are strikingly rigid. Only gradually is a broader approach to a cultural history of sensual (including visual) perception beginning to replace stories of technical progress, prehistories of the cinema, and defenses of disciplinary terrains—science against culture, scholars against amateurs, epistemology against entertainment, what is certified in the image against

what is fleeting and uncertain. My sketch offers just a few waymarks in a field that is still very underresearched, following the lead of Laurent Mannoni, *The Great Art of Light and Shadow: Archaeology of the Cinema* (Exeter: University of Exeter Press, 2000); Tom Gunning, "Phantasmagoria and the Manufacturing of Illusions and Wonder: Towards a Cultural Optics of the Cinematic Apparatus," in André Gaudreault, Catherine Russell, and Pierre Veronneau, eds., *Le cinématographe, nouvelle technologie du XXe siècle = The Cinema, a New Technology for the 20th Century* (Lausanne: Payot, 2004), pp. 31–44; Barbara Maria Stafford, *Artful Science: Enlightenment Entertainment and the Eclipse of Visual Education* (Cambridge, MA: MIT Press, 1994); Barbara Maria Stafford and Frances Terpak, *Devices of Wonder: From the World in a Box to Images on a Screen* (Los Angeles: Getty Research Institute for the History of Art & the Humanities, 2001).

46. See Ratcliff, *Quest for the Invisible.*

47. On the history of projection apparatuses, see, for example, John H. Hammond, *The Camera Obscura: A Chronicle* (Bristol: Adam Hilger, 1981); Hermann Hecht, *Pre-Cinema History: An Encyclopaedia and Annotated Bibliography of the Moving Image before 1896* (London: Bowker Saur, 1993); Paul Liesegang, *Zahlen und Quellen zur Geschichte der Projektionskunst und Kinematographie* (Berlin: Deutsches Drucks- und Verlagshaus, 1926); Friedrich von Zglinicki, *Der Weg des Films: Die Geschichte der Kinematographie und ihrer Vorläufer* (Berlin: Privately published, 1956); Jonathan Crary, *Techniques of the Observer: On Vision and Modernity in the Nineteenth Century* (Cambridge, MA: MIT Press, 1991), ch. 2; Olaf Breidbach, Kerrin Klinger, and André Karliczek, eds., *Natur im Kasten: Lichtbild, Schattenriss, Umzeichnung und Naturselbstdruck um 1800* (Jena: Ernst-Haeckel-Haus, 2010); Deac Rossell, "The Magic Lantern and Moving Images before 1800," *Barockberichte* 40–41 (2005), pp. 686–93; Rossell, *Laterna magica — Magic Lantern* (Stuttgart: Füsslin, 2008). From a cultural historian's perspective, see also Simon During, *Modern Enchantments: The Cultural Power of Secular Magic* (Cambridge, MA: Harvard University Press, 2002).

48. On Kepler and the camera obscura, see "Kepler, Optical Imagery, and the Camera Obscura," ed. Alan Shapiro, special issue, *Early Science and Medicine* 13.3 (2008).

49. On the following points, see Mannoni, *Great Art of Light and Shadow*, pp. 6–13. In the early sixteenth century (Mannoni names the period between 1521 and 1550), the camera obscura was refined by replacing the simple opening with a biconvex lens, making the projected image considerably sharper. On Leonardo and Cardano, see ibid., pp. 6–7; on della Porta, pp. 8–10; on the walk-in camera obscura in Paris, p. 13, and Hammond, *Camera Obscura*, p. 29; on eighteenth-century Greenwich, Hammond, *Camera Obscura*, pp. 71–72.

50. On Leeuwenhoek, Vermeer, and their ways of seeing, see Robert D. Huerta, *Giants of Delft: Johannes Vermeer and the Natural Philosophers; The Parallel Search for Knowledge during the Age of Discovery* (Lewisburg, PA: Bucknell University Press, 2003), and Snyder, *Eye of the Beholder*, with comments on Leeuwenhoek's possible use of a kind of solar microscope at pp. 296–98. On Vermeer's and possibly Leeuwenhoek's use of the camera obscura, see Philip Steadman, *Vermeer's Camera: Uncovering the Truth behind the Masterpieces* (Oxford: Oxford University Press, 2001), pp. 44–58, esp. pp. 48–49.

51. On Huygens's *laterna magica*, see Mannoni, *Great Art of Light and Shadow*, pp. 34–45; Koen Vermeir, "The Magic of the Magic Lantern (1660–1700): On Analogical Demonstration and the Visualization of the Invisible," *British Journal for the History of Science* 38.2 (2005), pp. 127–59. Huygens was in contact with Leeuwenhoek and competed with him on the investigation of spermatozoa after 1677, building his own microscopes for the purpose. Huygens's microscopes have hitherto been studied in isolation from his familiarity with other uses of the lens, for example in Marian Fournier, "Huygens' Designs for a Simple Microscope," *Annals of Science* 46.6 (1989), pp. 575–96.

52. Robert Hooke, "A Contrivance to make the Picture of any thing appear on a Wall, Cub-board, or within a Picture-frame, &c. in the midst of a Light room in the day-time; or in the Night-time in any room that is enlightned with a considerable number of Candles; devised and communicated by the Ingenious Mr. Hook, as follows," *Philosophical Transactions of the Royal Society of London* 3.38 (1668), p. 742.

53. Ibid., p. 743. On Hooke's and the Royal Society's involvement with projection more generally, see Matthew C. Hunter, "'Mr. Hooke's Reflecting Box': Modeling the Projected Image in the Early Royal Society," *Huntington Library Quarterly* 78.2 (2015), pp. 301–28.

54. Jean Leurechon, *Récréations mathématiques, composés de plusieurs problemes, plaisans & facetieux, d'arithmetique; geometrie, astrologie, optique, perspective, mechanique, chymie, & d'autres rares & curieux secrets: plusieurs desquels n'ont jamais esté imprimez* (Rouen: Charles Osmont, 1629), pp. 5–6. See Mannoni, *Great Art of Light and Shadow*, pp. 12–13; Hecht, *Pre-Cinema History*, p. 11.

55. Jean-François Nicéron, *La perspective curieuse* (Paris: Langlois, 1652), pp. 21–22. See Mannoni, *Great Art of Light and Shadow*, pp. 12–13; Hecht, *Pre-Cinema History*, pp. 13 and 17.

56. On regional differences within Europe and distribution beyond Europe, see Mannoni, *Great Art of Light and Shadow*, ch. 3.

57. On the microscopic tradition in Germany, especially southern Germany, see also

Ratcliff, *Quest for the Invisible*, pp. 24–25. On production of magic lanterns in Nuremberg, see Deac Rossell, "Nürnberg and the Bull's-Eye Magic Lantern," *New Magic Lantern Journal* 9.5 (2003), pp. 71–75.

58. Joannes Franciscus Griendelius, *Micrographia nova: oder Neu-curieuse Beschreibung verschiedener kleiner Cörper, welche vermittelst eines absonderlichen von dem Author neuerfundenen Vergrösser-Glases verwunderlich gross vorgestellet werden* (Nuremberg: Johann Zieger, 1687). The best-known account of Griendel's arts of projection and of seventeenth-century projection more generally is that of a traveler: the travel account published by the exiled Parisian physician Charles Patin (1633–1693) in 1674 impressively described a visit to Griendel's phantasmagoric show in Nuremberg. Charles Patin, *Travels thro' Germany, Bohemia, Swisserland, Holland, and other parts of Europe* (London: A. Swall and T. Child, 1696), pp. 232–36. On Griendel, see Rossell, "Magic Lantern." The fourteen engravings of microscopic objects in Griendel's *Micrographia nova* do not match up to his own skills as a performer or indeed to other contemporary drawings. See Fournier, *Fabric of Life*, p. 153. They are, moreover, purely static representations.

59. Johann Christoph Sturm, *Collegium experimentale sive Curiosum* (Nuremberg: Endter, 1676), pp. 161–66.

60. Johann Zahn, *Oculus artificialis teledioptricus sive telescopium*, 2nd ed. (Nuremberg: Lochner, 1702), pp. 730–31. On Zahn, see Mannoni, *Great Art of Light and Shadow*, pp. 63–66.

61. Zahn, *Oculus artificialis*, p. 731.

62. Ibid., pp. 730–31.

63. Barbara Maria Stafford, *Body Criticism: Imaging the Unseen in Enlightenment Art and Medicine* (Cambridge, MA: MIT Press, 1991), p. 360.

64. On Enlightenment science, see William Clark, Jan Golinski, and Simon Schaffer, eds., *The Sciences in Enlightened Europe* (Chicago: University of Chicago Press, 1999); Roy Porter, ed., *Eighteenth-Century Science* (Cambridge: Cambridge University Press, 2003); G. S. Rousseau and Roy Porter, eds., *The Ferment of Knowledge: Studies in the Historiography of Eighteenth-Century Science* (Cambridge: Cambridge University Press, 1980); Ulrich Johannes Schneider, ed., *Kulturen des Wissens im 18. Jahrhundert* (Berlin: De Gruyter, 2008); Emma C. Spary, *Utopia's Garden: French Natural History from Old Regime to Revolution* (Chicago: University of Chicago Press, 2000); and Stafford, *Artful Science*.

65. See Jim Bennett, "The Social History of the Microscope," *Journal of Microscopy* 155.3 (1989), pp. 267–80.

66. On the spectacular aspect, see Stafford, *Body Criticism*; Stafford, *Artful Science*;

and Stafford and Terpak, *Devices of Wonder*. On the discursive aspect, see Peter Heering, "The Enlightened Microscope: Re-enactment and Analysis of Projections with Eighteenth-Century Solar Microscopes," *British Journal for the History of Science* 41.3 (2008), pp. 345–67, describing reenactments of solar microscopy performances; also see Ratcliff, *Quest for the Invisible*.

67. Heering, "The Enlightened Microscope." Heering counts eighty-three examples in the large European collections and museums alone. Marc Ratcliff lists the instrument makers offering microscopes in their sales catalogues. Ratcliff, *Quest for the Invisible*, p. 150 n.8 (on the European market for microscopes, also chs. 1 and 4).

68. Martin Frobenius Ledermüller, *Ledermüllers . . . mikroskopische Gemüths- und Augen-Ergötzung: Bestehend, in Ein Hundert nach der Natur gezeichneten und mit Farben erleuchteten Kupfertafeln, Sammt deren Erklärung* (Nuremberg: Christian de Launoy, 1761), with comments on the history of the solar microscope on pp. 40–46; August Johann Rösel von Rosenhof, *Der monatlich herausgegebenen Insecten-Belustigung* (Nuremberg: Privately published, 1736–1752); Wilhelm Friedrich Freiherr von Gleichen genannt Rußworm, *Das Neueste aus dem Reiche der Pflanzen, oder Mikroskopische Untersuchungen und Beobachtungen der geheimen Zeugungstheile der Pflanzen in ihren Blüten, und der in denselben befindlichen Insekten . . .* (Nuremberg: s.n., 1764). Also Henry Baker, *The Microscope Made Easy: or, I. The Nature, Uses, and Magnifying Powers of the Best Kinds of Microscopes . . . II. An Account of What Surprising Discoveries Have Already Been Made by the Microscope*, 2nd ed. (London: Printed for R. Dodsley, 1743), pp. 22–26; Benjamin Martin, *The Young Gentleman and Lady's Philosophy: In a Continued Survey of the Works of Nature and Art by Way of Dialogue* (London: W. Owen, 1759), pp. 182–97; J. H. A. Dunker, *Mikroskopische Blätter oder Beschreibungen und vergrößernde Abbildungen der kleinsten Werke Gottes: Zum Nutzen und Vergnügen für erwachsene Kinder und ungeübte Beobachter* (Brandenburg: Leichsche Buchhandlung, 1798), pp. 3–7; Jean-Antoine Nollet, *Leçons de physique expérimentale*, 3rd ed. (Paris: Guerin & Delatour, 1764), vol. 5, pp. 567–79. On Zahn and the dissemination of microscopy books in Europe, and opposing the assertion of a general decline of microscopy in the period 1690–1710, see the empirical analysis in Ratcliff, *Quest for the Invisible*, esp. ch. 1.

69. See Heidrun Ludwig, *Nürnberger naturgeschichtliche Malerei im 17. und 18. Jahrhundert* (Marburg an der Lahn: Basilisken-Presse, 1998); Wilhelm Schwemmer, *Nürnberger Kunst im 18. Jahrhundert* (Nuremberg: Edelmann, 1974); Claus Nissen, *Die botanische Buchillustration*, 3 vols. (Stuttgart: Hiersemann, 1966); Thomas Schnalke, *Natur im Bild: Anatomie und Botanik in der Sammlung des Nürnberger Arztes Christoph Jacob Trew*, exhibition

catalogue (Erlangen: Universitätsverlag, 1995); Andreas Kraus, "Der Beitrag Frankens zur Entwicklung der Wissenschaften (1550–1800)," in Andreas Kraus, ed., *Handbuch der Bayerischen Geschichte*, vol. 3.1: *Geschichte Frankens bis zum Ausgang des 18. Jahrhunderts* (Munich: C. H. Beck, 1971), pp. 1054–1108; Gunter Mann, "Medizinisch-naturwissenschaftliche Buchillustration im 18. Jahrhundert in Deutschland," *Marburger Sitzungsberichte* 86.1–2 (1964), pp. 3–48; Richard Wegner, "Christoph Jacob Trew (1695–1769): Ein Führer zur Blütezeit naturwissenschaftlicher Abbildungswerke in Nürnberg im 18. Jahrhundert," *Mitteilungen zur Geschichte der Medizin, der Naturwissenschaften und der Technik* 39 (1940), pp. 218–28.

70. Against this interpretation, Ratcliff identifies a new visual regime of sequentiality and temporalization in the eighteenth century, based on the discoveries of the regeneration of polyps and the division of infusoria. Ratcliff, "Temporality." Countering Ratcliff, see Janina Wellmann, "Metamorphosis in Images: Insect Transformation from the End of the Seventeenth to the Beginning of the Nineteenth Century," in Gemma Anderson and John Dupré, eds., *Drawing Processes of Life: Molecules, Cells, Organisms* (Bristol: Intellect Press, 2022), pp. 237–71, and Wellmann, *The Form of Becoming: Embryology and the Epistemology of Rhythm, 1760–1830*, trans. Kate Sturge (New York: Zone Books, 2017).

71. Angela Fischel, "Optik und Utopie: Mikroskopische Bilder als Argument im 18. Jahrhundert," in Horst Bredekamp and Pablo Schneider, eds., *Visuelle Argumentationen: Die Mysterien der Repräsentation und die Berechenbarkeit der Welt* (Munich: Fink, 2006), pp. 257 and 259. Scholarship on Ledermüller's and Rösel's works has hitherto focused almost entirely on the copper engravings and the large illustrative plates accompanying the books. See Fischel, "Optik und Utopie"; Angela Fischel, ed., *Instrumente des Sehens* (Berlin: De Gruyter, 2004); Friedrich Klemm, *Martin Frobenius Ledermüller: Aus der Zeit der Salon-Mikroskopie des Rokoko* (Schweidnitz: B. Köhn, 1928).

72. Elkins, "On Visual Desperation"; see also Stafford, "Images of Ambiguity."

73. Stafford, *Body Criticism*, p. 362.

74. On the courtly context of Ledermüller's work, his post as inspector of the natural history collection belonging to his patron Margrave Friedrich of Bayreuth, the printing of his microscopical works, and his solar microscope demonstrations in Erlangen and at the Bayreuth court, see Gerhard H. Müller, "Martin Frobenius Ledermüllers Beziehungen zum Bayreuther Hof unter Markgraf Friedrich," *Jahrbuch für Fränkische Landesforschung* 38 (1978), pp. 171–80; Emil Reicke, ed., *Gottscheds Briefwechsel mit dem Nürnberger Naturforscher Martin Frobenius Ledermüller und dessen seltsame Lebensschicksale* (Leipzig:

Scholtze, 1923), pp. 19–22 and 149–57. In a letter to Johann Christoph Gottsched, Leder-müller describes a visit from von Gleichen-Rußworm: "His Excellency Privy Councilor von Gleichen also came quite unexpectedly from Bonnland to see me in Erlangen and not only did me the honor of his own presence, but also brought with him a party of the most refined persons to watch my observations in the darkened chamber with the solar microscope. In particular, I had the honor of welcoming His Excellency Privy Councilor von Rothkirch and his lady wife and other chevaliers and ladies." In Reicke, *Gottscheds Briefwechsel*, p. 131, and on experiments in the garden, p. 59.

75. On Ledermüller, see Karl Wilhelm Naumann, *Ledermüller und von Gleichen-Russ-worm: Zwei deutsche Mikroskopisten der Zopfzeit* (Leipzig: Scholtze Nachf., 1926); Klemm, *Martin Frobenius Ledermüller*; Reicke, *Gottscheds Briefwechsel*; Marc J. Ratcliff, *Genèse d'une découverte: La division des infusoires (1765–1766)* (Paris: Museum National Histoire Naturelle, 2016); Ratcliff, *Quest for the Invisible*, pp. 184–88.

76. Martin Frobenius Ledermüller, *Physicalische Beobachtungen derer Saamenthiergens, durch die allerbesten Vergrößerungs-Gläser und bequemlichsten Microscope betrachtet; und mit einer unpartheyischen Untersuchung und Gegeneinanderhaltung derer Buffonischen und Leu-wenhoeckischen Experimenten in einem Sendschreiben mit denen hierzu gehörigen Figuren und Kupfern einem Liebhaber der Natur-Kunde und Warheit mitgetheilet* (Nuremberg: George Peter Monath, 1756), p. 21. The ell as a historical unit of measurement is derived from the length of the forearm. Although this measure varied considerably, Ledermüller's explanation gives an impression of the projection's great size.

77. E. T. A. Hoffmann, "Master Flea" (1822), in *The Golden Pot and Other Tales,* trans. with an introduction and notes by Ritchie Robertson (Oxford: Oxford University Press, 1992), pp. 261–62.

78. On optical instruments in the work of E. T. A. Hoffmann, see Ulrich Stadler, "Von Brillen, Lorgnetten, Fernrohren und Kuffischen Sonnenmikroskopen: Zum Gebrauch optischer Instrumente in Hoffmanns Erzählungen," in Hartmut Steinecke, ed., *E.T.A. Hoffmann: Deutsche Romantik im europäischen Kontext* (Berlin: Erich Schmidt, 1993), pp. 91–105.

79. Ledermüller, *Mikroskopische Gemüths- und Augen-Ergötzung*, p. 145.

80. Ibid., p. 76.

81. Ibid., p. 146.

82. Ibid., p. 48.

83. Ibid., p. 169.

NOTES TO PAGES 120-122

84. A long passage is devoted to the study of the skin. Ibid., pp. 105 and 107.

85. See Heering, "Enlightened Microscope," pp. 363–64.

86. Buffon's theory of generation has been discussed extensively. See Roger, *Life Sciences*, pp. 439–52. On Buffon's concept of organic matter, see Peter Hanns Reill, *Vitalizing Nature in the Enlightenment* (Berkeley: University of California Press, 2005), ch. 1, esp. pp. 42–47. On the gender aspect of his theory, see Florence Vienne, "Organic Molecules, Parasites, 'Urthiere': The Contested Nature of Spermatic Animalcules, 1749–1841," trans. Kate Sturge, in Susanne Lettow, ed., *Gender, Race and Reproduction: Philosophy and the Early Life Sciences in Context* (Albany: SUNY Press, 2014), pp. 45–63. On the status of Buffon and his *Histoire naturelle* in eighteenth-century natural history, see Spary, *Utopia's Garden*, ch. 1.

87. Georges-Louis Leclerc Buffon, *Natural History: General and Particular*, trans. William Smellie, 2nd ed., 9 vols. (Edinburgh: William Creech, 1780–85), vol. 2, p. 221. There, Buffon addresses Leeuwenhoek's observations at length. On the idea that these are not animals but organic "particles" or "molecules," see ibid., pp. 190–92, 214, 220, 236; on the motion of the animalcules, pp. 200 and 228.

88. On the history of sperm, see Francis Joseph Cole, *Early Theories of Sexual Generation* (Oxford: Clarendon Press of Oxford University Press, 1930); Roger, *Life Sciences*; Vienne, "Organic Molecules."

89. Ledermüller, *Physicalische Beobachtungen*, p. 21.

90. Ibid., p. 11, also p. 22. See also Martin Frobenius Ledermüller, *Versuch einer gründlichen Vertheidigung derer Saamengethiergen: nebst einer kurzen Beschreibung der Leeuwenhoeckischen Mikroskopien und einem Entwurf zu einer vollständigern Geschichte des Sonnenmikroskops als der besten Rechtfertigung der Leeuwenhoeckischen Beobachtungen* (Nuremberg: George Peter Monath, 1758), p. 21.

91. Ledermüller, *Physicalische Beobachtungen*, pp. 12 and 11. Ledermüller himself could see "living creatures" in sperm "for no longer than two hours." Ibid., p. 15.

92. Martin Frobenius Ledermüller, *Nachlese seiner mikroskopischen Gemüths- und Augenergötzung; 1. Sammlung. Bestehend in zehn illuminierten Kupfertafeln, sammt deren Erklärung; und einer getreuen Anweisung, wie man alle Arten Mikroskope, geschickt, leicht und nüzlich gebrauchen solle* (Nuremberg: Christian de Launoy, 1762). Conversely, he deduces that Leeuwenhoek must already have used one. See also Ledermüller, *Versuch einer gründlichen Vertheidigung*, pp. 45–46.

93. On von Gleichen-Rußworm's life, see Melchior A. Weikard, *Biographie des Herrn*

Wilhelm Friedrich v. Gleichen genannt Rußworm Herrn auf Greifenstein, Bonnland und Ezelbach (Frankfurt am Main: Hermann, 1783).

94. Friedrich Wilhelm von Gleichen-Rußworm, *Abhandlung ueber die Samen- und Infusionstierchen und ueber die Erzeugung: Nebst mikroskopischen Beobachtungen des Samens der Tiere und verschiedener Infusionen* (Nuremberg: Winterschmidt, 1778), p. 6; also pp. 10 and 68.

95. Lazzaro Spallanzani, *Opusculi di fisica animale, e vegetabile. Volume secondo* (Modena: La Società Tipografica, 1776). This and the following quotations are from the English translation: Lazaro Spallanzani, *Tracts on the Nature of Animals and Vegetables*, trans. John Graham Dalyell (Edinburgh: William Creech, 1799), p. 133. On Spallanzani in general, see Giuseppe Montalenti et al., eds., *Lazzaro Spallanzani e la biologia del Settecento: Teorie, esperimenti, istituzioni scientifiche* (Florence: Olschki, 1982).

96. Spallanzani, *Tracts*, pp. 71–72 and 108.

97. Ibid., p. 109. On Spallanzani's dispute with Linnaeus, see pp. 104–14.

98. Ibid., pp. 110 and 130–32.

99. Ibid., pp. 77 and 89.

100. Ibid., pp. 135–47.

101. Ibid., pp. 39–41.

102. Ibid., p. 104.

103. Carolus Linnaeus, *Systema naturae*, 12th ed., 3 vols. (Stockholm: Lars Salvius, 1767), vol. 1.2, pp. 1326–27. Later in the eighteenth century, the Danish microscopist Otto Frederik Müller would classify hundreds of infusoria. Müller, *Animalcula infusoria fluviatilia et marina* (Copenhagen: Nicolai Möller, 1786). On naming and classifying microscopic animals of all kinds, see Ratcliff, *Quest for the Invisible*, ch. 8, esp. pp. 188–89, 209, 212.

104. Von Gleichen-Rußworm, *Abhandlung*, p. 98.

105. Ibid., p. 6.

106. Ibid., p. 98.

107. Ibid., p. 99.

108. Louis Joblot, *Descriptions et usages de plusieurs nouveaux microscopes tant simples que composez, avec de nouvelles observations faites sur une multitude innombrable d'insectes et d'autres animaux de diverses espèces qui naissent dans les liqueurs préparées et dans celles qui ne le sont point* (Paris: Jacques Collombat, 1718). On Joblot, see Hubert Lechevalier, "Louis Joblot and his Microscopes," *Bacteriological Reviews* 40.1 (1976), pp. 241–58; Ratcliff, *Quest for the Invisible*, ch. 2.

109. Joblot, *Descriptions et usages*, p. 13.

110. Ibid., pp. 21–22 and 29.

111. Ibid., p. 35.

112. Ibid., pp. 35 and 50.

113. Ibid., p. 56.

114. See, for example, Johann Friedrich Wilhelm Koch, *Mikrographie* (Magdeburg: Keil, 1803), vol. 1, pp. viii–xiii, who lists the items required for microscopy: "a good magnifying glass…a handheld microscope…a compound lens…a solar microscope in a dark room…numerous wooden sliders…glass slides with small indentations ground into them…a frog plate…search-glass…magnifying glass on a ball-joint arm…tweezers…hair brush…sharpened quills…capillary tubes…a soft chamois cloth to clean the lenses." The most comprehensive history of eighteenth-century microscopy is Ratcliff, *Quest for the Invisible*. See also Stafford, "Images of Ambiguity." On amateurs, see Jacques Roger, "The Living World," in Porter and Rousseau, eds., *The Ferment of Knowledge*, pp. 255–82. On the shipping of living specimens and Trembley's experiments, see Ratcliff, *Quest for the Invisible*, ch. 5.

115. On the division of infusoria, see Ratcliff, *Genèse d'une découverte*. On Trembley, see Aram Vartanian, "Trembley's Polyp, La Mettrie, and Eighteenth-Century Materialism," *Journal of the History of Ideas* 11.3 (1950), pp. 259–86; Virginia P. Dawson, *Nature's Enigma: The Problem of the Polyp in the Letters of Bonnet, Trembley and Réaumur* (Philadelphia, PA: American Philosophical Society, 1987); Marino Buscaglia, "La pratique, la figure et les mots dans les *Mémoires* d'Abraham Trembley sur les polypes (1744), comme exemple de communication scientifique," in Massimo Galuzzi, Gianni Micheli, and Maria Teresa Monti, eds., *Le forme della comunicazione scientifica* (Milan: Franco Angeli, 1998), pp. 313–46. On regeneration, see Charles E. Dinsmore, ed., *A History of Regeneration Research: Milestones in the Evolution of a Science* (Cambridge: Cambridge University Press, 1991). On spontaneous generation, see Farley, *Spontaneous Generation Controversy*; Lippmann, *Urzeugung*; and Paula Gottdenker, "Three Clerics in Pursuit of 'Little Animals,'" *Clio Medica* 14.3-4 (1980), pp. 213–24.

CHAPTER SIX: FROM ANIMAL TO ACTIVITY

1. See Antoni van Leeuwenhoek, "Observations, communicated to the Publisher by Mr. Antony van Leewenhoeck, in a Dutch Letter of the 9th of Octob. 1676, here English'd: Concerning little Animals by him observed in Rain- Well- Sea- and Snow-water; as also

in water wherein Pepper had lain infused," *Philosophical Transactions of the Royal Society of London* 12.133 (1677), pp. 829 and 830; also Leeuwenhoek, "XVI. A Letter from Mr. Anthouy van Leeuwenhoek, F. R. S. containing some further Microscopical Observations on the Animalcula found upon Duckweed, &c.," *Philosophical Transactions of the Royal Society of London* 28.337 (1713), pp. 160–64.

2. August Johann Rösel von Rosenhof, *Der monathlich-herausgegebenen Insecten-Belustigung dritter Theil* . . . (Nuremberg: Fleischmann, 1755), p. 617.

3. Ibid., p. 620. Rösel is not quite sure about this mechanism. In fact, the green algae he describes move by turning about their longitudinal axis.

4. Ibid., pp. 619 and 618.

5. Ibid., p. 623.

6. Félix Dujardin, "Recherches sur les organismes inférieurs," *Annales des Sciences naturelles (zoologie)* 2, sér. 4 (1835), pp. 343 and 361.

7. Christian Gottfried Ehrenberg, *Die Infusionsthierchen als vollkommene Organismen: Ein Blick in das tiefere organische Leben der Natur* (Leipzig: Leopold Voss, 1838), p. ***. As well as to the Rotifera (Rotatoria), Ehrenberg classified the infusoria as belonging to the "Polygastrica." He derived this name from his incorrect supposition that the food vacuoles often to be seen in infusoria were stomachs and digestive systems. Ehrenberg included in the Polygastrica essentially all of today's protozoa. See Armin Geus, "Christian Gottfried Ehrenberg und 'Die Infusionsthierchen als vollkommene Organismen: Ein Blick in das tiefere organische Leben der Natur' (1838)," *Medizinhistorisches Journal* 22.2–3 (1987), esp. p. 240; Frederick B. Churchill, "The Guts of the Matter: Infusoria from Ehrenberg to Bütschli, 1838–1876," *Journal of the History of Biology* 22.2 (1989), pp. 190–92; Otto Bütschli, *Infusoria und System der Radiolaria*, vol. 1.3 of *H. G. Bronns Klassen und Ordnungen des Tierreichs, wissenschaftlich dargestellt in Wort und Bild* (Leipzig: Winter'sche Verlagshandlung, 1887–89), pp. 1139–46. A contradictory stance is found in Félix Dujardin, *Histoire naturelle des zoophytes: Infusoires, comprenant la physiologie et la classification de ces animaux, et la manière de les étudier à l'aide du microscope* (Paris: Roret, 1841).

8. Dujardin, "Recherches," p. 354; Ehrenberg, *Die Infusionsthierchen*, p. v.

9. Frederick B. Churchill, "Introduction: Toward the History of Protozoology," *Journal of the History of Biology* 22.2 (1989), p. 185; Churchill calls his 1989 special issue "the first modern attempt to probe the intellectual and even institutional dimensions of protozoology" (p. 186). The classic historical surveys, with special attention to medical bacteriology, include Hubert A. Lechevalier and Morris Solotorovsky, *Three Centuries*

of Microbiology (New York: McGraw-Hill, 1965); and Patrick Collard, *The Development of Microbiology* (Cambridge: Cambridge University Press, 1976). See also Milton Wainwright, "An Alternative View of the Early History of Microbiology," *Advances in Applied Microbiology* 52 (2003), pp. 333–55.

10. Churchill, "The Guts of the Matter," p. 204.

11. The cell theory marked a watershed in the taxonomy of microorganisms. Detailed accounts of the numerous attempts to classify infusoria, from the first efforts by John Hill and Otto Frederik Müller in the eighteenth century up to the authors' own day, are given by Ehrenberg, *Die Infusionsthierchen*, Dujardin, *Histoire naturelle des zoophytes*, and Bütschli, *Infusoria und System der Radiolaria*. A comprehensive, if dated survey of the literature in a still constantly developing field is John O. Corliss, *The Ciliated Protozoa: Characterization, Classification and Guide to the Literature*, 2nd ed. (Oxford: Pergamon, 1979).

12. Georg August Goldfuss, *Ueber die Entwicklungsstufen des Thieres: Omne vivum ex ovo* (Nuremberg: Leonhard Schrag, 1817), pp. 21–22, also pp. 25, 32, 34, and diagram. Goldfuss himself categorized as protozoa not only the infusoria, lithozoa, and phytozoa, but also the medusae.

13. C. T. E. von Siebold, *Lehrbuch der vergleichenden Anatomie der wirbellosen Thiere* (Berlin: Veit, 1848), p. 3. On Siebold, see Bütschli, *Infusoria und System der Radiolaria*, pp. 1154–55.

14. Otto Bütschli even asserted "an almost unanimous rejection of the unicellular theory among infusoria researchers." Bütschli, *Infusoria und System der Radiolaria*, p. 1180. On the few dissenting voices, such as Thomas Huxley, Martin Barry, and Albert von Kölliker, see ibid., pp. 1153, 1155, 1181. On Kölliker and Siebold, see Churchill, "The Guts of the Matter," pp. 193–96. On Bütschli himself, see Richard Goldschmidt, "Otto Bütschli 1848–1920," *Die Naturwissenschaften* 8.28 (1920), pp. 543–49.

15. The third part of this study is dedicated to conjugation in the infusoria and also surveys the diversity of contemporary research opinions on the sexual reproduction and reproductive organs of the infusoria. Otto Bütschli, *Studien über die ersten Entwicklungsvorgänge der Eizelle, die Zelltheilung und die Conjugation der Infusorien* (Frankfurt am Main: C. Winter, 1876); Bütschli, *Infusoria und System der Radiolaria*; see also Churchill, "The Guts of the Matter."

16. Bütschli, *Studien über die ersten Entwicklungsvorgänge*, p. 159; see also Churchill, "The Guts of the Matter."

17. Bütschli, *Infusoria und System der Radiolaria.*

18. Natasha X. Jacobs, "From Unit to Unity: Protozoology, Cell Theory, and the New Concept of Life," *Journal of the History of Biology* 22.2 (1989), p. 218.

19. Friedrich Kützing, *Über die Verwandlung der Infusorien in niedere Algenformen* (Nordhausen: W. Köhne, 1844), p. v.

20. On the differences between the taxonomies of Owen, Hogg, Wilson, and Haeckel, and especially their different evolutionary implications, see Lynn J. Rothschild, "Protozoa, Protista, Protoctista: What's in a Name?," *Journal of the History of Biology* 22.2 (1989), pp. 277–305. Unlike the others, Haeckel, in *Generelle Morphologie der Organismen: Allgemeine Grundzüge der organischen Formenwissenschaft, mechanisch begründet durch die von Charles Darwin reformirte Deszendenztheorie* (Berlin: Georg Reimer, 1866), does not yet include infusoria in his Protista. Although Haeckel had made a name for himself in the study of single-celled organisms with an 1862 publication on radiolarians, it was only in 1873 that he pronounced infusoria to be single-cell organisms as well, classifying them as such in his *Das Protistenreich* of 1878. See Haeckel, *Das Protistenreich: Eine populäre Uebersicht über das Formengebiet der niedersten Lebewesen. Mit einem wissenschaftlichen Anhange: System der Protisten* (Leipzig: Ernst Günther, 1878).

21. Edmund B. Wilson, *The Cell in Development and Inheritance* (New York: Macmillan, 1896), p. 13; Laurence Picken, *The Organization of Cells and Other Organisms* (Oxford: Clarendon Press of Oxford University Press, 1960), p. 3; Andrew S. Reynolds, *The Third Lens: Metaphor and the Creation of Modern Cell Biology* (Chicago: University of Chicago Press, 2018), p. 18. On modern critiques of the cell theory, see also Andrew Reynolds, "The Redoubtable Cell," *Studies in History and Philosophy of Science Part C: Studies in History and Philosophy of Biological and Biomedical Sciences* 41.3 (2010), pp. 194–201.

22. Reynolds distinguishes four hypotheses of the classical cell theory: the "structural-anatomical thesis," in which cells are the building blocks of plants; the "developmental-physiological thesis"; the "evolutionary-phylogenetic thesis," propounded especially by Ernst Haeckel in the second half of the nineteenth century; and, toward the end of the century, a "genetic thesis of heredity." Added to that, after 1850, was "the aspect of a distinctively social theory." Reynolds, *The Third Lens*, pp. 23 and 25. Of the extensive literature on the cell theory, see Henry Harris, *The Birth of the Cell* (New Haven, CT: Yale University Press, 1999); John R. Baker, "The Cell-Theory: A Restatement, History, and Critique. Part I," *Quarterly Journal of Microscopic Science* 89.1 (1949), pp. 103–25; Baker, "The Cell-Theory: A Restatement, History, and Critique. Part II," *Quarterly Journal of*

Microscopic Science 90.1 (1949), pp. 87–108; Baker, "The Cell-Theory: A Restatement, History, and Critique. Part III: The Cell as a Morphological Unit," *Quarterly Journal of Microscopic Science* 93.2 (1952), pp. 157–90; Baker, "The Cell-Theory: A Restatement, History, and Critique. Part IV: The Multiplication of Cells," *Quarterly Journal of Microscopic Science* 94.4 (1953), pp. 407–40; Baker, "The Cell-Theory: A Restatement, History, and Critique. Part V: The Multiplication of Nuclei," *Quarterly Journal of Microscopic Science* 96.4 (1955), pp. 449–81; François Duchesneau, *Genèse de la théorie cellulaire*, Collections analytiques (Montreal: Bellarmin, 1987); more broadly, Karl S. Matlin, Jane Maienschein, and Manfred D. Laubichler, eds., *Visions of Cell Biology: Reflections Inspired by Cowdry's General Cytology* (Chicago: University of Chicago Press, 2018).

23. Matthias Jacob Schleiden, "Beiträge zur Phytogenesis," *Archiv für Anatomie, Physiologie und wissenschaftliche Medicin* (1838), p. 138. The larger epistemic question of how to think about units and unity, isolation and integration, remains relevant to this day. Historically, the question has always been accompanied by political and social metaphors—from Johann Christian Reil's eighteenth-century republic of the bodily organs to Rudolf Virchow's democratic "cell state" a hundred years later. On Virchow, Herbert Spencer, and Ernst Haeckel, see, for example, Reynolds, *The Third Lens*, esp. pp. 25–30; Kathrin Sander, *Organismus als Zellenstaat: Rudolf Virchows Körper-Staat-Metapher zwischen Medizin und Politik* (Herbolzheim: Centaurus, 2012); James Elwick, "Herbert Spencer and the Disunity of the Social Organism," *History of Science* 41.1 (2003), pp. 35–72; and more generally, Jan Sapp, *Evolution by Association: A History of Symbiosis* (New York: Oxford University Press, 1994), esp. ch. 3.

24. Theodor Schwann, *Microscopical Researches into the Accordance in the Structure and Growth of Animals and Plants*, trans. Henry Smith (London: Sydenham Society, 1847), pp. 165–66.

25. Ibid., p. 192.

26. Reynolds, *The Third Lens*, pp. 23 and 22.

27. See Thomas Steele Hall, *Ideas of Life and Matter: Studies in the History of General Physiology; 600 B.C.–1900 A.D.*, 2 vols. (Chicago: University of Chicago Press, 1969), vol. 2; Harris, *The Birth of the Cell*; Daniel Liu, "The Cell and Protoplasm as Container, Object, and Substance, 1835–1861," *Journal of the History of Biology* 50.4 (2017), pp. 889–925; Eugenio Frixione, "Recurring Views on the Structure and Function of the Cytoskeleton: A 300-Year Epic," *Cell Motility* 46.2 (2000), pp. 73–94; John V. Pickstone, "Globules and Coagula: Concepts of Tissue Formation in the Early Nineteenth Century," *Journal of the History*

of Medicine and Allied Sciences 28.4 (1973), pp. 336–56; Ohad Parnes, "The Envisioning of Cells," *Science in Context* 13.1 (2000), pp. 71–92; Baker, "The Cell-Theory. Part I."

28. Pickstone, "Globules and Coagula," p. 336; Parnes, "Envisioning of Cells," p. 74.

29. What is today called cytoplasmic streaming. Bonaventura Corti, *Osservazioni microscopiche sulla tremella e sulla circolazione del fluido in una pianta acquajuola* (Lucca: Giuseppe Rocchi, 1774); Ludolf Christian Treviranus, *Beyträge zur Pflanzenphysiologie* (Göttingen: Heinrich Dieterich, 1811).

30. See Janina Wellmann, *The Form of Becoming: Embryology and the Epistemology of Rhythm, 1760–1830*, trans. Kate Sturge (New York: Zone Books, 2017).

31. Jean-Baptiste Lamarck, *Zoological Philosophy: An Exposition with Regard to the Natural History of Animals*, trans. Hugh Elliott (London: Macmillan, 1914), pp. 183 and 188.

32. Ibid., p. 189. The French word *gangue* has the geological meaning of a vein or seam running through rock or can mean a hard or sclerotic envelope of tissue. Lamarck's use of the word indicates a conjunction of motion and petrifaction, the organization of organic matter out of movement.

33. Ibid.

34. Ibid., pp. 186, 187, 226. On vitalism, see Chapter 1.

35. Everett Mendelsohn, "Cell Theory and the Development of General Physiology," *Archives internationales d'histoire des sciences* 16 (1963), p. 424. On Schwann's cell theory as part of a larger research program of explaining vital processes by specific causal agents, see Parnes, "Envisioning of Cells."

36. On Dutrochet, see Duchesneau, *Genèse de la théorie cellulaire*, pp. 21–45; Harris, *The Birth of the Cell*, pp. 27–31; Hall, *Ideas of Life and Matter*, vol. 2, pp. 184–88; Joseph Schiller and Tetty Schiller, *Henri Dutrochet (Henri du Trochet 1776–1847): Le matérialisme mécaniste et la physiologie générale* (Paris: Albert Blanchard, 1975).

37. Henri Dutrochet, *Recherches anatomiques et physiologiques sur la structure intime des animaux et des végétaux, et sur leur motilité* (Paris: Baillière, 1824), p. 6.

38. Ibid., p. 163.

39. Henri Dutrochet, *L'agent immédiat du mouvement vital dévoilé dans sa nature et dans son mode d'action chez les végétaux et chez les animaux* (Paris: Baillière, 1826), p. v.

40. Dutrochet, *Recherches anatomiques*, p. 5. See also Dutrochet, *L'agent immédiat*.

41. Dutrochet, *Recherches anatomiques*, pp. 5–6.

42. On Dutrochet's studies of osmosis and his application of physics to physiology, see John V. Pickstone, "Vital Actions and Organic Physics: Henri Dutrochet and French

Physiology during the 1820s," *Bulletin of the History of Medicine* 50.2 (1976), pp. 191–212; Jacques Bolard, "L'osmose et la vie selon Dutrochet," *BibNum* (2012), https://journals. openedition.org/bibnum/508.

43. Dutrochet, *L'agent immédiat*, p. vi.

44. Ibid., p. 115.

45. Henri Dutrochet, "Nouvelles observations sur l'endosmose et l'exosmose, et sur la cause de ce double phénomène," *Annales de chimie et de physique* 35 (1827), p. 393.

46. Dutrochet, *L'agent immédiat*, p. 89.

47. Ibid., p. 225.

48. In physical terms, osmosis means the passage of solvent molecules through a semipermeable membrane against the concentration gradient; to equalize the concentrations, the flow is from the less concentrated to the more concentrated solution. Dutrochet realized that the membrane is necessary to this process and also that the phenomenon is not restricted to organic membranes. He did not, however, identify the principle of semipermeability. See Dutrochet, "Nouvelles observations," p. 393.

49. Dutrochet, *L'agent immédiat*, pp. 190 and 191.

50. Ibid., p. 140.

51. Dutrochet, *Recherches anatomiques*, pp. 203 and 204.

52. Dutrochet, *L'agent immédiat*, pp. 200–201.

53. Dutrochet, "Nouvelles observations," p. 400.

54. Schiller and Schiller, *Henri Dutrochet*, p. 31. Brownian motion, named for the English botanist Robert Brown, was observed at around the same time (1828), first in plants, then in inorganic substances. It was interpreted as a physical movement of particles and thus not necessarily organic or animal in character. On Brownian motion, see Stephen G. Brush, "A History of Random Processes: I. Brownian Movement from Brown to Perrin," *Archive for History of Exact Sciences* 5.1 (1968), pp. 1–36.

55. "Le double phénomène de l'endosmose et de l'exosmose est, *par le fait* et non *par sa nature* un phénomène exclusivement physiologique." Dutrochet, "Nouvelles observations," p. 400 (original emphasis).

56. Dutrochet, *L'agent immédiat*, p. 55.

57. On the development of French microscopy in the early to mid-nineteenth century, see Ann La Berge, "Medical Microscopy in Paris, 1830–1855," in Ann La Berge and Mordechai Feingold, eds., *French Medical Culture in the Nineteenth Century* (Leiden: Brill, 1994), pp. 296–326. Dutrochet studied in Paris, like François Magendie, who from the

1830s held the first chair of experimental physiology at the Collège de France and was one of the key figures of this new school of thought in France. On visual explanation and the role of projection techniques and living specimens in late nineteenth-century German physiology, see Henning Schmidgen, "Cinematography without Film: Architecture and Technologies of Visual Instruction in Biology around 1900," in Nancy Anderson and Michael R. Dietrich, eds., *The Educated Eye: Visual Culture and Pedagogy in the Life Sciences* (Hanover, NH: Dartmouth College Press, 2012), pp. 94–120.

58. Dutrochet, *L'agent immédiat*, pp. 58–59.

59. Ibid., p. 59. Later studies showed Dutrochet that he had been mistaken; the movements described by Schultz-Schultzenstein do indeed exist. Interestingly, Dutrochet now found the evidence for this in the rhythmic alternation of motion and stasis. Ibid., pp. 60–70.

60. Dujardin, *Histoire naturelle des zoophytes*, p. 189.

61. See Louis Joubin, "Notices biographiques, X: Félix Dujardin," *Archives de parasitologie* 4.1 (1901), pp. 5–57. Dujardin's natural history of infusoria contains a long passage on microscopy. Dujardin, *Histoire naturelle des zoophytes*, pp. 189–206. In 1843, he published a much-read manual on the use of the microscope: Félix Dujardin, *Nouveau manuel complet de l'observateur au microscope* (Paris: Librairie encyclopédique de Roret, 1843).

62. Dujardin created the new class of Rhizopoda for these infusoria, based on the formless substance of which the creatures consisted and their form of motion—by expanding their body mass. Dujardin, *Histoire naturelle des zoophytes*, pp. 240–59, 122. He counted as infusoria organisms that are today classified as amoebae, foraminifera, and flagellates. Apart from the flagellates, these all move by forming pseudopodia.

63. Dujardin, "Recherches," p. 354.

64. Ibid., p. 356.

65. Ibid., pp. 359–60.

66. Dujardin, *Histoire naturelle des zoophytes*, p. 202.

67. Dujardin, "Recherches," p. 367. On Dujardin's research and his notion of the *sarcode*, see E. Fauré-Fremiet, "L'oeuvre de Félix Dujardin et la notion de protoplasma," *Protoplasma* 23.1 (1935), pp. 250–69; Hall, *Ideas of Life and Matter*, vol. 2, pp. 171–78.

68. Dujardin, *Histoire naturelle des zoophytes*, pp. 37–38. As mentioned, Ehrenberg believed these vacuoles were stomachs.

69. Ibid., p. 26; see also p. 42.

70. Albrecht von Haller regarded contractility as a characteristic of muscles alone;

see Chapter 1. Later, for Bichat, every vital response was a kind of contraction. Hall, *Ideas of Life and Matter*, vol. 2, pp. 121–32.

71. Siebold, *Lehrbuch der vergleichenden Anatomie*, p. 8. The term *parenchyma* was used to describe soft plant tissue by Nehemiah Grew in *Anatomy of Plants* (1682), where he compared the parenchyma to the threads in bobbin lace.

72. Alexander Ecker, *Zur Lehre vom Bau und Leben der contractilen Substanz der niedersten Thiere* (Basel: Schweighauser, 1846), p. 3.

73. Ibid., p. 15; see also p. 19.

74. Ibid., p. 18.

75. Ibid., p. 22. Like Ecker with regard to polyps, Haeckel describes contractility (in this case, of the sarcode) as the "cause of all the phenomena of movement" among radiolarians. Haeckel, *Die Radiolarien: Eine Monographie* (Berlin: Georg Reimer, 1862), p. 91.

76. Hugo von Mohl, "Über die Saftbewegung im Innern der Zelle," *Botanische Zeitung* 4 (1846), p. 73. On the membrane, see ibid., pp. 77 and 91–92.

77. Ibid., p. 75. The term *protoplasm* (literally "the first thing formed"), religious in origin, was applied as early as 1839 by Jan Purkyně (1787–1869), though neither to infusoria nor to plants, but to young animal embryos. See Baker, "The Cell-Theory. Part II," 90–91. On the role of Purkyně in histology more broadly, see Harris, *The Birth of the Cell*, pp. 82–93. Mohl's cell structures are the nucleus and what he calls the "primordial utricle" (*Primordialschlauch*); see Liu, "Cell and Protoplasm."

78. Franz Unger, *Anatomie und Physiologie der Pflanzen* (Pest: C. A. Hartleben, 1855), pp. 281, 284, 280; see also Unger, *Botanische Briefe* (Vienna: Carl Gerold, 1852), p. 152.

79. Ferdinand Cohn, *Zur Naturgeschichte des Protococcus pluvialis Kützing (Haematococcus pluvialis Flotow, Chlamidococcus versatilis A. Braun, Chlamidococcus pluvialis Flotow u. A. Braun)* (Bonn: Eduard Weber, 1850), p. 663.

80. Anton de Bary, "Die Mycetozoen: Ein Beitrag zur Kenntnis der niedersten Thiere," *Zeitschrift für wissenschaftliche Zoologie* 10.1 (1859), p. 163.

81. Cohn, *Zur Naturgeschichte*, p. 663. The sparse literature on Cohn mainly deals with his microbiological work. See Gerhart Drews, "The Roots of Microbiology and the Influence of Ferdinand Cohn on Microbiology of the 19th Century," *FEMS Microbiology Reviews* 24.3 (2000), pp. 25–49; Brigitte Hoppe, "Die Biologie der Mikroorganismen von F. J. Cohn (1828–1898): Entwicklung aus Forschungen über mikroskopische Pflanzen und Tiere," *Sudhoffs Archiv* 67.2 (1983), pp. 158–89. His larger oeuvre is addressed by Margot

Klemm, *Ferdinand Julius Cohn 1828–1898: Pflanzenphysiologe, Mikrobiologe, Begründer der Bakteriologie* (Frankfurt am Main: Lang, 2003).

82. Cohn, *Zur Naturgeschichte*, pp. 665 and 663.

83. Ferdinand Cohn, *Contractile Gewebe im Pflanzenreiche* (Breslau: Josef Max, 1861), p. 48.

84. Max Johann Sigismund Schultze, "Über Muskelkörperchen und das was man eine Zelle zu nennen habe," *Archiv für Anatomie, Physiologie und wissenschaftliche Medicin* (1861), p. 9, also pp. 11, 13, 21.

85. On the concept of protoplasm, see the older surveys by George L. Goodale, "Protoplasm and Its History," *Botanical Gazette* 14.10 (1889), pp. 235–46; Edwin G. Conklin, "Cell and Protoplasm Concepts: Historical Account," in Forest Ray Moulton, ed., *The Cell and Protoplasm* (Washington, DC: The Science Press, 1940), pp. 6–19. See also Gerald L. Geison, "The Protoplasmic Theory of Life and the Vitalist-Mechanist Debate," *Isis* 60.3 (1969), pp. 273–92; Andrew Reynolds, "Amoebae as Exemplary Cells: The Protean Nature of an Elementary Organism," *Journal of the History of Biology* 41 (2008), pp. 307–37; Liu, "Cell and Protoplasm."

86. Schultze, "Über Muskelkörperchen," p. 12.

87. Ibid., p. 17.

88. From the correspondence of Ferdinand Cohn, quoted in Hoppe, "Die Biologie der Mikroorganismen," p. 169.

89. Max Johann Sigismund Schultze, *Das Protoplasma der Rhizopoden und der Pflanzenzellen: Ein Beitrag zur Theorie der Zelle* (Leipzig: Wilhelm Engelmann, 1863), pp. 11–27.

CHAPTER SEVEN: THE STUDY OF FORMATION

1. Uri Alon used this expression to voice his astonishment on first encountering the world of biology: "When I first read a biology textbook, it was like reading a thriller. Every page brought a new shock. As a physicist, I was used to studying matter that obeys precise mathematical laws. But cells are matter that dances." Alon, *An Introduction to Systems Biology: Design Principles of Biological Circuits* (Boca Raton, FL: Taylor & Francis, 2006), p. 1.

2. See Janina Wellmann, *The Form of Becoming: Embryology and the Epistemology of Rhythm, 1760–1830*, trans. Kate Sturge (New York: Zone Books, 2017).

3. Gräper published extensively on embryology throughout his life and was an internationally recognized researcher (he was one of the few members of the Institut

d'embryologie, Utrecht), but he never held a major post in Jena. Gräper himself believed that a hostile academic environment was responsible for his lack of academic recognition: the Jena institute, founded by Carl Gegenbaur, had a tradition of comparative anatomy and morphology, rather than of the experimental approaches favored by Gräper. In the Nazi-dominated environment of the 1930s, however, Gräper seems to have felt less excluded. More research remains to be done on this point, but it is already clear that from 1932 to 1938, the Anatomical Institute in Jena was headed by a member of the NSDAP and SS, Hans Böker (1886–1939). On the University of Jena in the Nazi period, see Uwe Hossfeld, ed., *"Kämpferische Wissenschaft": Studien zur Universität Jena im Nationalsozialismus* (Cologne: Böhlau, 2003); Hossfeld, *"Im Dienst an Volk und Vaterland": Die Jenaer Universität der NS-Zeit* (Cologne: Böhlau, 2005). Early biographers noted that Gräper was a member of the SA, the paramilitary Stahlhelm organization, and the German Society for Military Policy and Military Science (DGWW): "Ludwig Ernst Gräper," in Bernd Wiederanders and Susanne Zimmermann, eds., *Buch der Docenten der Medicinischen Facultät zu Jena* (Golmsdorf: Jenzig, 2004), pp. 101–104; Karl Peter, "Ludwig Gräper," *Anatomischer Anzeiger* (1937), p. 304. Nevertheless, his personnel file at the university does not include any entries in the rubric "Political activities (if a member of the NSDAP, membership number)." Ludwig Gräper, personnel file, Thüringisches Ministerium für Volksbildung, Universitätsarchiv Jena, Bestand D, no. 945.

4. Ludwig Gräper, "Untersuchungen über die Herzbildung der Vögel," *Archiv für Entwicklungsmechanik* 24 (1907), p. 378.

5. Ibid., p. 382.

6. Ibid., pp. 377–78. On staining techniques, see the still-reprinted handbook Benno Romeis, *Taschenbuch der mikroskopischen Technik* (Berlin: De Gruyter, 1943); also Jutta Schickore, "Fixierung mikroskopischer Beobachtungen: Zeichnung, Dauerpräparat, Mikrofotografie," in Peter Geimer, ed., *Ordnungen der Sichtbarkeit: Fotografie in Wissenschaft, Kunst und Technologie* (Frankfurt am Main: Suhrkamp, 2002), pp. 285–310. On microtechnology and the production of microscopy samples in general, see Brian Bracegirdle, *A History of Microtechnique: The Evolution of the Microtome and the Development of Tissue Preparation* (London: Heinemann, 1978).

7. Gräper, "Untersuchungen über die Herzbildung der Vögel," pp. 384 and 385.

8. Ludwig Gräper, "Beobachtung von Wachstumsvorgängen an Reihenaufnahmen lebender Hühnerembryonen nebst Bemerkungen über vitale Färbung," *Morphologisches Jahrbuch* 39.3-4 (1911), p. 304.

9. Gräper demonstrated that the embryo's death was the result not of injecting the stain, but of a nutritional disturbance caused by the neutral red. Gräper, "Beobachtung von Wachstumsvorgängen," p. 307. A contemporary source on vital staining is Leonor Michaelis, "Die vitale Färbung, eine Darstellungsmethode der Zellgranula," *Archiv für mikroskopische Anatomie* 55.1 (1899), pp. 558–75.

10. Gräper, "Beobachtung von Wachstumsvorgängen," pp. 305–306.

11. Ibid., p. 312.

12. Ibid., pp. 310–12.

13. Ibid., pp. 312–15.

14. Ibid., pp. 320–21; see also Ludwig Gräper, "Die Primitiventwicklung des Hühnchens nach stereokinematographischen Untersuchungen, kontrolliert durch vitale Farbmarkierung und verglichen mit der Entwicklung anderer Wirbeltiere," *Wilhelm Roux' Archiv für Entwicklungsmechanik* 116 (1929), p. 387.

15. Gräper, "Beobachtung von Wachstumsvorgängen," pp. 321–24. The two lines at the bottom of the diagram are already part of the second figure, fig. 4a.

16. Jean-Claude Beetschen, "Amphibian Gastrulation: History and Evolution of a 125 Year-Old Concept," *International Journal of Developmental Biology* 45 (2001), p. 781.

17. August Rauber and Friedrich Kopsch, *Lehrbuch und Atlas der Anatomie des Menschen* (Stuttgart: Thieme, 1870 et seq.). Both Gräper and Kopsch were dedicated teachers, though as yet we know little of their many students and their careers. On Kopsch's biography, see Wolfram Richter, "Friedrich Kopsch als Histologe und Embryologe: Zur Erinnerung an den großen Berliner Anatomen aus Anlaß seines 30. Todestages am 24. Januar 1985," *Zeitschrift für mikroskopisch-anatomische Forschung* 99.1 (1985), pp. 1–13; Anton Waldeyer, "Friedrich Kopsch," *Zeitschrift für mikroskopisch-anatomische Forschung* 61 (1955), pp. 155–58.

18. Friedrich Kopsch, "Ueber die Zellen-Bewegungen während des Gastrulationsprocesses an den Eiern vom Axolotl und vom braunen Grasfrosch," *Gesellschaft naturforschender Freunde Berlin* (1895), p. 21.

19. Friedrich Kopsch, "Beiträge zur Gastrulation beim Axolotl- und Froschei," *Verhandlungen der Anatomischen Gesellschaft Basel* 9 (1895), p. 182; Kopsch, "Ueber die Zellen-Bewegungen," p. 23.

20. Kopsch, "Beiträge zur Gastrulation," p. 182; Kopsch, "Ueber die Zellen-Bewegungen," p. 23.

21. Kopsch, "Beiträge zur Gastrulation," p. 182.

22. Kopsch, "Ueber die Zellen-Bewegungen," pp. 24-25.

23. Kopsch, "Beiträge zur Gastrulation," p. 182; Kopsch, "Ueber die Zellen-Bewegungen," p. 24.

24. Kopsch, "Beiträge zur Gastrulation," p. 183; Kopsch, "Ueber die Zellen-Bewegungen," pp. 24-25. For more on the work of Kopsch, see Janina Wellmann, "Model and Movement: Studying Cell Movement in Early Morphogenesis, 1900 to the Present," *History and Philosophy of the Life Sciences* 40.3 (2018), pp. 1-25.

25. Kopsch, "Ueber die Zellen-Bewegungen," pp. 29-30. The blastopore in amphibians is homologous to the primitive groove in birds.

26. Ludwig Gräper, "Die frühe Entwicklung des Hühnchens nach Kinoaufnahmen des lebenden Embryo," *Anatomischer Anzeiger* 61 (1926), pp. 55 and 57.

27. Ibid., pp. 54-56.

28. Gräper, "Primitiventwicklung nach stereokinematographischen Untersuchungen," p. 383.

29. Ludwig Gräper, "Die Primitiventwicklung des Hühnchens, verglichen mit der anderer Wirbeltiere, mit stereokinematographischen Demonstration," *Verhandlungen der Anatomischen Gesellschaft* (1928), p. 90, also pp. 93-94. In contrast, early films on morphogenesis, such as sea urchin development, did not explore new territory, but presented well-established knowledge. On Ries, see Julius Ries, "Kinematographie der Befruchtung und Zellteilung," *Archiv für mikroskopische Anatomie* 74 (1909), pp. 1-31; Hannah Landecker and Christopher Kelty, "A Theory of Animation: Cells, L-systems, and Film," *Grey Room* 17 (2004), pp. 30-63; Janina Wellmann, "Plastilin und Kreisel, Pinsel und Projektor: Julius Ries und die Materialität der seriellen Anschauung," in Gerhard Scholtz, ed., *Serie und Serialität: Konzepte und Analysen in Gestaltung und Wissenschaft* (Berlin: Reimer, 2017), pp. 77-93. On Vlès and Chevreton, see Louise Chevreton and Frederic Vlès, "La cinématique de la segmentation de l'œuf et la chronophotographie du développement de l'oursin," *Comptes rendus hebdomadaires des séances de l'Académie des sciences* 149 (1909), pp. 806-809. Early film has attracted much attention from historians of film, culture, and science alike. See, for example, Lisa Cartwright, *Screening the Body: Tracing Medicine's Visual Culture* (Minneapolis: University of Minnesota Press, 1995); Scott Curtis, *The Shape of Spectatorship: Art, Science, and Early Cinema in Germany* (New York: Columbia University Press, 2015); Mary Ann Doane, *The Emergence of Cinematic Time: Modernity, Contingency, and the Archive* (Cambridge, MA: Harvard University Press, 2002); Tom Gunning, "The Cinema of Attractions: Early Film, Its Spectator and the

Avant-Garde," in Thomas Elsaesser, ed., *Early Cinema: Space, Frame, Narrative* (London: BFI Publishing, 1990), pp. 56–62; Annemone Ligensa and Klaus Kreimeier, eds., *Film 1900: Technology, Perception, Culture* (New Barnet, UK: John Libbey, 2009). The similarities and distinctions between educational, scientific, and popular science films in the period has been discussed widely. See Oliver Gaycken, *Devices of Curiosity: Early Cinema and Popular Science* (Oxford: Oxford University Press, 2015); Gaycken, "The Swarming of Life: Moving Images, Education, and Views through the Microscope," *Science in Context* 24.3 (2011), pp. 361–80. On Jean Comandon, see Béatrice de Pastre and Thierry Lefebvre, *Filmer la science, comprendre la vie: Le cinéma de Jean Comandon* (Paris: CNC, 2012), and on the staging of cells and bacteria as drama, see Hannah Landecker, "Creeping, Drinking, Dying: The Cinematic Portal and the Microscopic World of the Twentieth-Century Cell," *Science in Context* 24.3 (2011), p. 399.

30. Gräper, "Primitiventwicklung des Hühnchens," pp. 90–91.

31. Gräper, "Primitiventwicklung nach stereokinematographischen Untersuchungen," pp. 387 and 383.

32. Gräper, "Die frühe Entwicklung des Hühnchens," p. 54. See also Friedrich Stier, "Siedentopf, Henry Friedrich Wilhelm," *Neue Deutsche Biographie* 24 (2010), www. deutsche-biographie.de/pnd117629251.html#ndbcontent.

33. Gräper, "Die frühe Entwicklung des Hühnchens," p. 54.

34. Gräper, "Primitiventwicklung nach stereokinematographischen Untersuchungen," p. 383. The film appears to have been made as early as 1925. Ibid., p. 415.

35. Gräper, "Die frühe Entwicklung des Hühnchens," p. 56.

36. Ibid., pp. 55–56.

37. Ludwig Gräper, "Die Methodik der stereokinematographischen Untersuchung des lebenden vitalgefärbten Hühnerembryos," *W. Roux' Archiv für Entwicklungsmechanik* 115 (1929), p. 523.

38. Anthony R. Michaelis, *Research Films in Biology, Anthropology, Psychology, and Medicine* (New York: Academic Press, 1955), p. 58.

39. Our understanding of spatial vision goes back only to the mid-nineteenth century. In 1838, Charles Wheatstone calculated and drew stereoscopic image pairs and constructed a viewing device, the stereoscope, in which the spectator's gaze is deflected onto the part images by mirrors. The first double-lens camera was made by the Scottish physicist David Brewster in 1849. Stereocinematography experienced a boom around 1900, but was soon superseded by film. On stereocinematography, see Paul Liesegang, *Zahlen*

und Quellen zur Geschichte der Projektionskunst und Kinematographie (Berlin: Deutsches Drucks- und Verlagshaus, 1926), pp. 106–109; Liesegang, *Wissenschaftliche Kinematographie: Einschließlich der Reihenphotographie* (Düsseldorf: Liesegang, 1920), pp. 190–211. On Gräper's stereocinematographic research, see Liesegang, *Wissenschaftliche Kinematographie*, p. 288; Michaelis, *Research Films*, pp. 31–32, 58, 116. Gräper himself gives a detailed account in Gräper, "Die Methodik der stereokinematographischen Untersuchung."

40. Gräper, "Die Methodik der stereokinematographischen Untersuchung," pp. 526–40.

41. Ibid., pp. 534–35.

42. Gräper, "Primitiventwicklung des Hühnchens," p. 90.

43. Gräper, "Primitiventwicklung nach stereokinematographischen Untersuchungen," fig. 8 and p. 406.

44. Ibid., pp. 401 and 391.

45. Ibid., p. 398. Gräper identifies the same countercurrent motion in amphibians. Ibid., pp. 399–400.

46. Ibid., p. 391.

47. Gräper, "Primitiventwicklung des Hühnchens," p. 91; Gräper, "Primitiventwicklung nach stereokinematographischen Untersuchungen," pp. 396–97.

48. Gräper, "Primitiventwicklung nach stereokinematographischen Untersuchungen," pp. 387–88.

49. Gräper, "Primitiventwicklung des Hühnchens," p. 93.

50. Ibid., p. 94.

51. Gräper, "Primitiventwicklung nach stereokinematographischen Untersuchungen," p. 409.

52. Ibid., pp. 420–23, quotation p. 420.

53. Verena Niethammer, "Indoktrination oder Innovation? Der Unterrichtsfilm als neues Lehrmedium im Nationalsozialismus," *Journal of Educational Media, Memory, and Society* 81 (2016), pp. 30–60; Malte Ewert, *Die Reichsanstalt für Film und Bild in Wissenschaft und Unterricht (1934–1945)* (Hamburg: Kovač, 1998). Explaining the role of the Reich Office, Gräper himself wrote that it supplied the unexposed film, copied the results, and recopied them onto cine film for use in teaching. Ludwig Gräper, "Die Gastrulation nach Zeitaufnahmen linear markierter Hühnerkeime," *Verhandlungen der Anatomischen Gesellschaft* (1937), p. 7.

54. Gräper, "Die Gastrulation nach Zeitaufnahmen," p. 6. See also Christian Reiß,

"Shooting Chicken Embryos: The Making of Ludwig Gräper's Embryological Films 1911–1940," *Isis* 112.2 (2021), pp. 299–306.

55. See Gräper, "Die Gastrulation nach Zeitaufnahmen," and the explanatory booklet: Hildegard Treiber-Merbach, *Gastrulation der Hühnerkeimscheibe: Erläuterungen zu dem gleichnamigen Film von Prof. Dr. med. Ludwig Gräper* (Berlin: Reichsanstalt für Film und Bild in Wissenschaft und Unterricht, 1939).

56. Treiber-Merbach, *Gastrulation der Hühnerkeimscheibe*, p. 4; Gräper, "Die Gastrulation nach Zeitaufnahmen," p. 10.

57. Gräper, "Primitiventwicklung nach stereokinematographischen Untersuchungen," p. 408; see also pp. 394, 396, 407, 414.

58. The work of G. Frommolt and Willi Kuhl and their institutions remain to be researched. G. Frommolt, "Die Befruchtung und Furchung des Kanincheneies im Film," *Zentralblatt für Gynäkologie* 58.1 (1934), pp. 7–12; Frommolt, *Befruchtung, Furchung und erste Teilungen des Kanincheneies: Hochschulfilm C 23 des Instituts für Film und Bild* (Berlin: Reichsstelle für den Unterrichtsfilm, 1936); Willi Kuhl and Hans Freska, *Die Entwicklung des Eies der weißen Maus* (Berlin: Reichsanstalt für Film und Bild, 1938). On the period up to 1931 and the medical context, see Alexander Friedland, "'...doch erscheint in seiner Denkschrift die Bedeutung des klinischen Films für den Unterricht allzustark betont': Zur Geschichte des Medizinisch-kinematographischen Instituts der Charité 1923–1931," *Medizinhistorisches Journal* 52.2–3 (2017), pp. 148–79.

59. These included Ronald Canti (1883–1936), a London pathologist and almost exact contemporary of Gräper's, at the Strangeways Research Laboratory in Cambridge, UK; the embryologist Warren Lewis (1870–1964) at the Carnegie Institution in Washington; and the surgeon Alexis Carrel (1873–1944) at the Rockefeller Institute for Medical Research in New York. On Canti, see G. E. H. Foxon, "Early Biological Film — The Work of R. G. Canti," *University Vision* 15 (1976), pp. 5–13. Lewis received inspiration and firsthand instruction from Canti in Cambridge, putting it to use at his home institution in Washington by making time-lapse studies of the development of rabbit eggs and investigating endocytosis (he called it "pinocytosis," drinking by cells): the flow of substances in and out of the cell. On mid-twentieth-century cell cinematography, see Landecker, "Creeping, Drinking, Dying"; Brian Stramer and Graham A. Dunn, "Cells on Film — The Past and Future of Cinemicroscopy," *Journal of Cell Science* 128 (2015), pp. 9–13. For a survey of the use of film in various research domains, see Michaelis, *Research Films*, and (up to the 1920s) Liesegang, *Wissenschaftliche Kinematographie*; Liesegang, *Zahlen und Quellen*.

60. Tryggve Gustafson and Lewis Wolpert, "The Cellular Basis of Morphogenesis and Sea Urchin Development," *International Review of Cytology* 15 (1963), p. 141.

61. Ibid., p. 143.

62. Ibid., pp. 202 and 141. On their filming, see Wellmann, "Model and Movement."

63. Michael Abercrombie, "Concepts in Morphogenesis," *Proceedings of the Royal Society of London. Series B. Biological Sciences* 199.1136 (1977), pp. 337 and 342. Abercrombie's short historical sketch focuses on anglophone research, and he neither mentions nor seems to be aware of Gräper's and Kopsch's work.

64. Landecker, "Creeping, Drinking, Dying," p. 412; see also Hannah Landecker, *Culturing Life: How Cells Became Technologies* (Cambridge, MA: Harvard University Press, 2007).

65. Hannah Landecker, "The Life of Movement: From Microcinematography to Live-Cell Imaging," *Journal of Visual Culture* 11.3 (2012), p. 388.

66. Scott Curtis, "Die kinematographische Methode: Das 'bewegte Bild' und die Brownsche Bewegung," *Montage/av* 14.2 (2005), p. 25.

67. Evelyn Fox Keller, *Making Sense of Life: Explaining Biological Development with Models, Metaphors, and Machines* (Cambridge, MA: Harvard University Press, 2002), p. 218.

CHAPTER EIGHT: THE ROOT OF ALL EXISTENCE

1. Lisa Chong, Elizabeth Culotta, and Andrew Sugden, "On the Move," *Science* 288.5463 (2000), p. 79.

2. Natasha Myers, *Rendering Life Molecular: Models, Modelers, and Excitable Matter* (Durham, NC: Duke University Press, 2015), p. 236 (original emphasis).

3. John Dupré and Daniel J. Nicholson, "A Manifesto for a Processual Philosophy of Biology," in Daniel J. Nicholson and John Dupré, eds., *Everything Flows: Towards a Processual Philosophy of Biology* (Oxford: Oxford University Press, 2018), p. 3.

4. Daniel J. Nicholson, "Organisms ≠ Machines," *Studies in History and Philosophy of Science Part C: Studies in History and Philosophy of Biological and Biomedical Sciences* 44.4, part B (2013), pp. 669–78; Nicholson, "The Machine Conception of the Organism in Development and Evolution: A Critical Analysis," *Studies in History and Philosophy of Science Part C: Studies in History and Philosophy of Biological and Biomedical Sciences* 48, part B (2014), pp. 162–74.

5. Examples are Donna Haraway's "dirty ontology," Judith Butler's "performativity," Paul Rabinow's "biosociality," and Marilyn Strathern's notion of "merographic connections."

6. Mariam Fraser, Sarah Kember, and Celia Lury, "Inventive Life: Approaches to the New Vitalism," *Theory, Culture & Society* 22.1 (2005), pp. 1–14.

7. Isabelle Stengers, "Beyond Conversation: The Risks of Peace," in Catherine Keller and Anne Daniell, eds., *Process and Difference: Between Cosmological and Poststructuralist Postmodernisms* (Albany: SUNY Press, 2002), p. 248.

8. Fraser, Kember, and Lury, "Inventive Life," p. 5.

9. Other organelles, such as the mitochondria, were first made visible using dyes around the turn of the twentieth century, but it was only from the mid-twentieth century on, with the advent of centrifugation (which separates the components of the cell) and electron microscopy, that they could be analyzed in detail.

10. Wilhelm Kühne, *Untersuchungen über das Protoplasma und die Contractilität* (Leipzig: Engelmann, 1864), p. 1.

11. Ibid.

12. Albert Szent-Györgyi, ed., *Studies from the Institute of Medical Chemistry, University of Szeged*, vol. 1: *Myosin and Muscular Contraction* (Basel: Karger, 1942).

13. The sliding filament hypothesis was set out independently in two 1950s papers: Andrew F. Huxley and Rolf Niedergerke, "Structural Changes in Muscle During Contraction: Interference Microscopy of Living Muscle Fibres," *Nature* 173.4412 (1954), pp. 971–73, and Hugh Huxley and Jean Hanson, "Changes in the Cross-Striations of Muscle during Contraction and Stretch and Their Structural Interpretation," *Nature* 173.4412 (1954), pp. 973–76.

14. Elias Lazarides and Klaus Weber, "Actin Antibody: The Specific Visualization of Actin Filaments in Non-Muscle Cells," *Proceedings of the National Academy of Sciences* 71.6 (1974), pp. 2268–72; Thomas D. Pollard and Edward D. Korn, "Acanthamoeba Myosin: I. Isolation from Acanthamoeba Castellanii of an Enzyme Similar to Muscle Myosin," *Journal of Biological Chemistry* 248.13 (1973), pp. 4682–90.

15. Gary Borisy et al., "Microtubules: 50 Years on from the Discovery of Tubulin," *Nature Reviews: Molecular Cell Biology* 17.5 (2016), p. 323.

16. J. E. Heuser and M. W. Kirschner, "Filament Organization Revealed in Platinum Replicas of Freeze-Dried Cytoskeletons," *Journal of Cell Biology* 86.1 (1980), p. 212.

17. Ibid.

18. D. L. Taylor and Y. L. Wang, "Molecular Cytochemistry: Incorporation of Fluorescently Labeled Actin into Living Cells," *Proceedings of the National Academy of Sciences of the United States of America* 75.2 (1978), pp. 857–61.

19. Robert Day Allen, Nina Strömgren Allen, and Jeffrey L. Travis, "Video-Enhanced

Contrast, Differential Interference Contrast (AVEC-DIC) Microscopy: A New Method Capable of Analyzing Microtubule-Related Motility in the Reticulopodial Network of Allogromia Laticollaris," *Cell Motility* 1.3 (1981), p. 291; see also S. Inoué, "Video Image Processing Greatly Enhances Contrast, Quality, and Speed in Polarization-Based Microscopy," *Journal of Cell Biology* 89.2 (1981), pp. 346–56.

20. Tim Mitchison and Marc Kirschner, "Dynamic Instability of Microtubule Growth," *Nature* 312.5991 (1984), pp. 237–42. In-vitro visualization soon followed: Tetsuya Horio and Hirokazu Hotani, "Visualization of the Dynamic Instability of Individual Microtubules by Dark-Field Microscopy," *Nature* 321.6070 (1986), pp. 605–607. See Gary J. Brouhard, "Dynamic Instability 30 Years Later: Complexities in Microtubule Growth and Catastrophe," *Molecular Biology of the Cell* 26.7 (2015), pp. 1207–10.

21. Hans-Jörg Rheinberger, *Toward a History of Epistemic Things: Synthesizing Proteins in the Test Tube* (Stanford, CA: Stanford University Press, 1997), p. 21.

22. Ibid., p. 3.

23. Michael P. Sheetz and James A. Spudich, "Movement of Myosin-Coated Fluorescent Beads on Actin Cables *In Vitro*," *Nature* 303.5912 (1983), pp. 33.

24. James A. Spudich, Stephen J. Kron, and Michael P. Sheetz, "Movement of Myosin-Coated Beads on Oriented Filaments Reconstituted from Purified Actin," *Nature* 315.6020 (1985), pp. 584–86.

25. Robert Day Allen et al., "Fast Axonal Transport in Squid Giant Axon," *Science* 218.4577 (1982), pp. 1127–29.

26. Ronald D. Vale, "How Lucky Can One Be? A Perspective from a Young Scientist at the Right Place at the Right Time," *Nature Medicine* 18.10 (2012), pp. 1486–88.

27. Ronald D. Vale et al., "Movement of Organelles along Filaments Dissociated from the Axoplasm of the Squid Giant Axon," *Cell* 40.20 (1985), pp. 449–54.

28. Allen, Allen, and Travis, "Video-Enhanced Contrast," p. 296.

29. Ronald D. Vale et al., "Organelle, Bead, and Microtubule Translocations Promoted by Soluble Factors from the Squid Giant Axon," *Cell* 40.3 (1985), pp. 559–69.

30. Ronald D. Vale, Thomas S. Reese, and Michael P. Sheetz, "Identification of a Novel Force-Generating Protein, Kinesin, Involved in Microtubule-Based Motility," *Cell* 42.1 (1985), pp. 39–50; Vale, "How Lucky Can One Be?"

31. Sheetz and Spudich, "Movement of Myosin-Coated Fluorescent Beads," p. 33; Spudich, Kron, and Sheetz, "Movement of Myosin-Coated Beads on Oriented Filaments."

32. Ron Vale uses his own bodily movement to convey that difference in his video

"Molecular Motor Proteins," www.youtube.com/watch?v=9RUHJhskW00, starting at minute 20:00.

33. J. Howard, A. J. Hudspeth, and R. D. Vale, "Movement of Microtubules by Single Kinesin Molecules," *Nature* 342.6246 (1989), p. 156.

34. Vale, Reese, and Sheetz, "Identification of a Novel Force-Generating Protein." Today, the protein is known as kinesin-1 or conventional kinesin. A range of kinesin molecules has been found since then, and today, there is a whole family of kinesin proteins.

35. Vale et al., "Movement of Organelles," p. 39.

36. Sheetz and Spudich, "Movement of Myosin-Coated Fluorescent Beads," pp. 34 and 31.

37. Howard, Hudspeth, and Vale, "Movement of Microtubules," pp. 154–58.

38. Ibid., p. 154.

39. Karel Svoboda et al., "Direct Observation of Kinesin Stepping by Optical Trapping Interferometry," *Nature* 365.6448 (1993), pp. 721–22.

40. Howard, Hudspeth, and Vale, "Movement of Microtubules," p. 154.

41. Svoboda et al., "Direct Observation of Kinesin Stepping," pp. 721–27.

42. Ibid., p. 726.

43. Jonathon Howard, "One Giant Step for Kinesin," *Nature* 365.6448 (1993), p. 696.

44. D. D. Hackney, J. D. Levitt, and J. Suhan, "Kinesin Undergoes a 9 S to 6 S Conformational Transition," *Journal of Biological Chemistry* 267.12 (1992), p. 8700; see also Yoshie Harada et al., "Mechanochemical Coupling in Actomyosin Energy Transduction Studied by *In Vitro* Movement Assay," *Journal of Molecular Biology* 216.1 (1990), pp. 49–68; Howard, "One Giant Step for Kinesin," p. 697.

45. Jonathon Howard, "The Movement of Kinesin Along Microtubules," *Annual Review of Physiology* 58.1 (1996), p. 704.

46. Ibid., pp. 703–29.

47. Ibid., p. 707.

48. Ibid., pp. 703–29; Günther Woehlke and Manfred Schliwa, "Walking on Two Heads: The Many Talents of Kinesin," *Nature Reviews Molecular Cell Biology* 1.1 (2000), pp. 50–58. The model has recently been underpinned by a new imaging technology called FIONA (fluorescence imaging one nanometer accuracy), which makes it possible to image the motion of an individual kinesin head. Ahmet Yildiz et al., "Kinesin Walks Hand-Over-Hand," *Science* 303.5658 (2004), p. 676; Ahmet Yildiz and Paul R. Selvin, "Kinesin: Walking, Crawling or Sliding Along?" *Trends in Cell Biology* 15.2 (2005), pp. 112–20.

49. Yildiz and Selvin, "Kinesin: Walking, Crawling or Sliding Along?," p. 116; Adrian N. Fehr et al., "On the Origin of Kinesin Limping," *Biophysical Journal* 97.6 (2009), pp. 1663–70.

50. Manfred Schliwa, "Kinesin: Walking or Limping?" *Nature Cell Biology* 5.12 (2003), p. 1043.

51. Charles L. Asbury, "Kinesin: World's Tiniest Biped," *Current Opinion in Cell Biology* 17.1 (2005), p. 89; Yildiz and Selvin, "Kinesin: Walking, Crawling or Sliding Along?," p. 113.

52. Ronald D. Vale and Ronald A. Milligan, "The Way Things Move: Looking Under the Hood of Molecular Motor Proteins," *Science* 288.5463 (2000), p. 89.

53. Asbury, "Kinesin: World's Tiniest Biped," p. 89; Howard, "The Movement of Kinesin," p. 706; Schliwa, "Kinesin: Walking or Limping?," pp. 1043–44.

54. Thus in Dryden's translation. Ovid, *Ovid's Metamorphoses in Fifteen Books, Translated by the Most Eminent Hands* (London: Jacob Tonson, 1717), p. 4. On human gait, see Kurt Bayertz, *Der aufrechte Gang: Eine Geschichte des anthropologischen Denkens* (Munich: Beck, 2012).

55. Likewise, the human gait remains one of the greatest challenges in modern robotics. See, for example, Oussama Khatib et al., "Robotics-Based Synthesis of Human Motion," *Journal of Physiology-Paris* 103.3 (2009), pp. 211–19; Andrey Rudenko et al., "Human Motion Trajectory Prediction: A Survey," *International Journal of Robotics Research* 39.8 (2020), pp. 895–935.

56. Alain Viel and Robert A. Lue, animation by John Liebler and the scientific animation company XVIVO, "The Inner Life of the Cell," www.youtube.com/watch?v=B_zD3NxSsD8. The video is no longer available on YouTube. An edited three-and-a-half-minute version is available on Vimeo at https://vimeo.com/236991501.

57. Hoogenraad Lab, "A Day in the Life of a Motor Protein," www.youtube.com/watch?v=tMKlPDBRJiE.

58. Ibid., minute 3:16–3:19.

59. On animation and automation, see Jackie Stacey and Lucy Suchman, eds., "Animation and Automation — The Liveliness and Labours of Bodies and Machines," special issue, *Body & Society* 18.1 (2012). On science and animation, see Olivia Banner and Kirsten Ostherr, eds., "Science/Animation," special issue, *Discourse* 37.3 (2015). On cinema and animation, see André Gaudreault and Philippe Gauthier, eds., "Could Kinematography Be Animation and Animation Kinematography?," special issue, *animation: an*

interdisciplinary journal 6.2 (2011). On the processes of imaging and imaging of processes, see Bettina Papenburg, Liv Hausken, and Sigrid Schmitz, eds., "The Processes of Imaging/Imaging of Processes," special section, *Catalyst: Feminism, Theory, Technoscience* 4.2 (2018). And on science and documentary, see Joshua Malitsky and Oliver Gaycken, eds., "Science and Documentary," special issue, *Journal of Visual Culture* 11.3 (2012). See also the edited volumes Karen Beckman, ed., *Animating Film Theory* (Durham, NC: Duke University Press, 2014); Suzanne Buchan, ed., *Pervasive Animation* (New York: Routledge, 2013); and Scott Curtis, ed., *Animation* (New Brunswick, NJ: Rutgers University Press, 2019). Classic studies on animation are Donald Crafton, *Before Mickey: The Animated Film 1898–1928* (Cambridge, MA: MIT Press, 1984); Alan Cholodenko, ed., *The Illusion of Life: Essays on Animation* (Sydney: Power Publications, 1991); and Paul Wells, *Understanding Animation* (London: Routledge, 1998). On the broader context of animism, see Irene Albers and Anselm Franke, eds., *Animismus: Revisionen der Moderne* (Zurich: Diaphanes, 2012).

60. Myers, *Rendering Life Molecular*, p. 232.

61. Natasha Myers, "Animating Mechanism: Animations and the Propagation of Affect in the Lively Arts of Protein Modelling," *Science & Technology Studies* 19.2 (2006), p. 25.

62. Ibid., pp. 24 and 17 (original emphasis). See also Myers, *Rendering Life Molecular*.

63. Myers, *Rendering Life Molecular*, p. 232.

64. Alla Gadassik, "Assembling Movement: Scientific Motion Analysis and Studio Animation Practice," *Discourse* 37.3 (2015), p. 269. Gadassik shows how, in the early phase of cartoons—especially Walt Disney Studios in the 1930s to 1960s—shared technologies, analytic interests, and aesthetics shaped the animation industry and scientific approaches in equal measure.

65. Tsung-Li Liu et al., "Observing the Cell in Its Native State: Imaging Subcellular Dynamics in Multicellular Organisms," *Science* 360.6386 (2018), eaaq1392; Michael W. Davidson and Richard N. Day, eds., *The Fluorescent Protein Revolution* (Boca Raton, FL: CRC Press, 2014); Martin Chalfie, "GFP: Lighting Up Life," *Proceedings of the National Academy of Sciences* 106.25 (2009), pp. 10073–80; Yingxiao Wang, John Y. J. Shyy, and Shu Chien, "Fluorescence Proteins, Live-Cell Imaging, and Mechanobiology: Seeing Is Believing," *Annual Review of Biomedical Engineering* 10.1 (2008), pp. 1–38.

66. Hannah Landecker, "The Life of Movement: From Microcinematography to Live-Cell Imaging," *Journal of Visual Culture* 11.3 (2012), p. 383; see also Hannah Landecker and Christopher Kelty, "A Theory of Animation: Cells, l-Systems, and Film," *Grey Room* 17 (2004), pp. 30–63.

67. Landecker, "The Life of Movement," pp. 393 and 394. Landecker argues that live-cell imaging originated in the animation of a theory of life that was based on the gene, but increasingly shifted toward proteins because they, and not genes themselves, were what became visible through fluorescent labeling. The term "molecular vitalism" was coined by the Harvard systems biologist Marc Kirschner. Marc Kirschner, John Gerhart, and Tim Mitchison, "Molecular 'Vitalism,'" *Cell* 100.1 (2000), pp. 79–88.

68. Quoted in Edward S. Small and Eugene Levinson, "Toward a Theory of Animation," *Velvet Light Trap* 24 (1989), p. 68 (original emphasis).

69. Tom Gunning, "Animating the Instant: The Secret Symmetry between Animation and Photography," in Karen Beckman, ed., *Animating Film Theory* (Durham, NC: Duke University Press, 2014), p. 40.

70. Svoboda et al., "Direct Observation of Kinesin Stepping," p. 722.

71. See Hyokeun Park, Erdal Toprak, and Paul R. Selvin, "Single-Molecule Fluorescence to Study Molecular Motors," *Quarterly Reviews of Biophysics* 40.1 (2007), p. 87; Natalia Fili, "Single-Molecule and Single-Particle Imaging of Molecular Motors *In Vitro* and *In Vivo*," in Christopher P. Toseland and Natalia Fili, eds., *Fluorescent Methods for Molecular Motors* (Basel: Springer, 2014), pp. 141–59; William J. Greenleaf, Michael T. Woodside, and Steven M. Block, "High-Resolution, Single-Molecule Measurements of Biomolecular Motion," *Annual Review of Biophysics and Biomolecular Structure* 36.1 (2007), pp. 171–90; Dawen Cai et al., "Recording Single Motor Proteins in the Cytoplasm of Mammalian Cells," *Methods in Enzymology* 475 (2010), pp. 81–107; Miguel Coelho, Nicola Maghelli, and Iva M. Tolić-Nørrelykke, "Single-Molecule Imaging *In Vivo*: The Dancing Building Blocks of the Cell," *Integrative Biology* 5.5 (2013), pp. 748–58.

72. Ahmet Yildiz et al., "Myosin V Walks Hand-Over-Hand: Single Fluorophore Imaging with 1.5-nm Localization," *Science* 300.5628 (2003), pp. 2061–65; Yildiz et al., "Kinesin Walks Hand-Over-Hand" (2004), pp. 676–78; Yildiz and Paul R. Selvin, "Fluorescence Imaging with One Nanometer Accuracy: Application to Molecular Motors," *Accounts of Chemical Research* 38.7 (2005), pp. 574–82.

73. Till Korten et al., "Fluorescence Imaging of Single Kinesin Motors on Immobilized Microtubules," in Erwin J. G. Peterman and Gijs J. L. Wuite, eds., *Single Molecule Analysis: Methods and Protocols* (Totowa, NJ: Humana Press, 2011), pp. 121–37; Park, Toprak, and Selvin, "Single-Molecule Fluorescence"; Fili, "Single-Molecule and Single-Particle Imaging."

74. Hiroshi Isojima et al., "Direct Observation of Intermediate States during the Stepping Motion of Kinesin-1," *Nature Chemical Biology* 12.4 (2016), pp. 290–97.

75. Erwin J. G. Peterman, "Kinesin's Gait Captured," *Nature Chemical Biology* 12.4 (2016), p. 206.

76. Sébastien Courty et al., "Tracking Individual Kinesin Motors in Living Cells Using Single Quantum-Dot Imaging," *Nano Letters* 6.7 (2006), pp. 1491–95.

77. Tyler J. Chozinski, Lauren A. Gagnon, and Joshua C. Vaughan, "Twinkle, Twinkle Little Star: Photoswitchable Fluorophores for Super-Resolution Imaging," *FEBS Letters* 588.19 (2014), pp. 3603–12.

78. Maria Dienerowitz et al., "Single-Molecule FRET Dynamics of Molecular Motors in an ABEL Trap," *Methods* (online February 2021).

79. Chalfie, "GFP: Lighting Up Life."

80. Joel E. Cohen, "Mathematics Is Biology's Next Microscope, Only Better; Biology Is Mathematics' Next Physics, Only Better," *PLoS Biology* 2.12 (2004), p. 2017.

81. Ibid.

82. Ibid., pp. 2017–23.

83. Gunnar Johansson, "Rigidity, Stability, and Motion in Perceptual Space: A Discussion of Some Mathematical Principles Operative in the Perception of Relative Motion and of Their Possible Function as Determinants of a Static Perceptual Space," *Acta Psychologica* 14 (1958), p. 369. See Chapter 3.

84. Nicholas Rescher, *Process Philosophy: A Survey of Basic Issues* (Pittsburgh, PA: University of Pittsburgh Press, 2000), p. 22; also Johannes Jaeger and Nick Monk, "Everything Flows: A Process Perspective on Life," *EMBO Reports* 16.9 (2015), p. 1065.

85. Rescher, *Process Philosophy*, p.13.

86. Ibid.

87. Erin Manning, *Relationscapes: Movement, Art, Philosophy* (Cambridge, MA: MIT Press, 2009), pp. 14, 17, 19.

Works Cited

Abbott, Jake J., Kathrin E. Peyer, Marco Consentino Lagomarsino, Li Zhang, Lixin Dong, Ioannis K. Kaliakatsos, and Bradley J. Nelson. "How Should Microrobots Swim?" *International Journal of Robotics Research* 28.11–12 (2009), pp. 1434–47.

Abercrombie, Michael. "Concepts in Morphogenesis." *Proceedings of the Royal Society of London. Series B. Biological Sciences* 199.1136 (1977), pp. 337–44.

Adam, Hans Christian, ed. *Eadweard Muybridge: The Human and Animal Locomotion Photographs*. Cologne: Taschen, 2010.

Adelmann, Howard B. *Marcello Malpighi and the Evolution of Embryology*. 5 vols. Ithaca, NY: Cornell University Press, 1966.

Aguilar, Jeffrey, Tingnan Zhang, Feifei Qian, Mark Kingsbury, Benjamin McInroe, Nicole Mazouchova, Chen Li, et al. "A Review on Locomotion Robophysics: The Study of Movement at the Intersection of Robotics, Soft Matter, and Dynamical Systems." *Reports on Progress in Physics* 79.11 (2016), 110001.

Akin, Demir, Jennifer Sturgis, Kathy Ragheb, Debby Sherman, Kristin Burkholder, Paul J. Robinson, Arun K. Bhunia, Sulma Mohammed, and Rashid Bashir. "Bacteria-Mediated Delivery of Nanoparticles and Cargo into Cells." *Nature Nanotechnology* 2 (2007), pp. 441–49.

Alapan, Yunus, Oncay Yasa, Berk Yigit, I. Ceren Yasa, Pelin Erkoc, and Metin Sitti. "Microrobotics and Microorganisms: Biohybrid Autonomous Cellular Robots." *Annual Review of Control, Robotics, and Autonomous Systems* 2.1 (2019), pp. 205–30.

Albers, Irene, and Anselm Franke, eds. *Animismus: Revisionen der Moderne*. Zurich: Diaphanes, 2012.

Allen, Julia. *Swimming with Dr. Johnson and Mrs. Thrale: Sport, Health and Exercise in Eighteenth-Century England.* Cambridge: Lutterworth Press, 2012.

Allen, Robert Day, J. Metuzals, I. Tasaki, S. T. Brady, and S. P. Gilbert. "Fast Axonal Transport in Squid Giant Axon." *Science* 218.4577 (1982), pp. 1127–29.

——, Nina Strömgren Allen, and Jeffrey L. Travis. "Video-Enhanced Contrast, Differential Interference Contrast (AVEC-DIC) Microscopy: A New Method Capable of Analyzing Microtubule-Related Motility in the Reticulopodial Network of Allogromia Laticollaris." *Cell Motility* 1.3 (1981), pp. 291–302.

Alon, Uri. *An Introduction to Systems Biology: Design Principles of Biological Circuits.* Boca Raton, FL: Taylor & Francis, 2006.

Alpers, Svetlana. *The Art of Describing: Dutch Art in the Seventeenth Century.* Chicago: University of Chicago Press, 1983.

Amelunxen, Hubertus von, Dieter Appelt, and Michael Baumgartner, eds. *Notation: Kalkül und Form in den Künsten.* Berlin: Akademie der Künste Berlin, 2008.

Aristotle. *De anima.* Trans. J. A. Smith. In *The Basic Works of Aristotle.* Ed. with an introduction by Richard McKeon. New York: Random House, 1941.

——. *De motu animalium.* Trans. A. S. L. Farquharson. In *The Works of Aristotle.* Eds. J. A. Smith and W. D. Ross. Vol. 5. Oxford: Clarendon Press of Oxford University Press, 1912.

——. *De motu animalium: Über die Bewegung der Lebewesen, Griechisch-Deutsch.* Ed. Oliver Primavesi. Trans. Klaus Corcilius. Hamburg: Felix Meiner, 2018.

——. *Parva Naturalia with On the Motion of Animals.* Trans. David Bolotin. Macon, GA: Mercer University Press, 2021.

——. *Physics.* Trans. R. P. Hardie and R. K. Gaye. In *The Basic Works of Aristotle.* Ed. with an introduction by Richard McKeon. New York: Random House, 1941.

——. *The Physics.* 2 vols. Trans. Francis M. Cornford and Philip H. Wicksteed. Cambridge, MA: Harvard University Press, 1980.

Asbury, Charles L. "Kinesin: World's Tiniest Biped." *Current Opinion in Cell Biology* 17.1 (2005), pp. 89–97.

Baker, Henry. *The Microscope Made Easy: or, I. The Nature, Uses, and Magnifying Powers of the Best Kinds of Microscopes . . . II. An Account of What Surprising Discoveries Have Already Been Made by the Microscope.* 2nd ed. London: Printed for R. Dodsley, 1743.

Baker, John R. "The Cell-Theory: A Restatement, History, and Critique. Part I." *Quarterly Journal of Microscopic Science* 89.1 (1949), pp. 103–25.

——. "The Cell-Theory: A Restatement, History, and Critique. Part II." *Quarterly*

Journal of Microscopic Science 90.1 (1949), pp. 87–108.

_____. "The Cell-Theory: A Restatement, History, and Critique. Part III: The Cell as a Morphological Unit." *Quarterly Journal of Microscopic Science* 93.2 (1952), pp. 157–90.

_____. "The Cell-Theory: A Restatement, History, and Critique. Part IV: The Multiplication of Cells," *Quarterly Journal of Microscopic Science* 94.4 (1953), pp. 407–40.

_____. "The Cell-Theory: A Restatement, History, and Critique. Part V: The Multiplication of Nuclei." *Quarterly Journal of Microscopic Science* 96.4 (1955), pp. 449–81.

Baldini, Ugo. "Animal Motion before Borelli, 1600–1680." In Domenico Bertoloni Meli, ed., *Marcello Malpighi: Anatomist and Physician*, pp. 193–246. Florence: Olschki, 1997.

Balzac, Honoré de. *Comédie humaine*. 7 vols. New York: Century, 1911.

Banner, Olivia, and Kirsten Ostherr, eds. "Science/Animation." Special issue, *Discourse* 37.3 (2015).

Barbour, Julian B. *Absolute or Relative Motion? Volume 1, The Discovery of Dynamics. A Study from a Machian Point of View of the Discovery and the Structure of Dynamical Theories.* Cambridge: Cambridge University Press, 1989.

Bary, Anton de. "Die Mycetozoen: Ein Beitrag zur Kenntnis der niedersten Thiere." *Zeitschrift für wissenschaftliche Zoologie* 10.1 (1859), pp. 88–173.

Bayertz, Kurt. *Der aufrechte Gang: Eine Geschichte des anthropologischen Denkens*. Munich: C. H. Beck, 2012.

Beckman, Karen, ed. *Animating Film Theory*. Durham, NC: Duke University Press, 2014.

Beetschen, Jean-Claude. "Amphibian Gastrulation: History and Evolution of a 125 Year-Old Concept." *International Journal of Developmental Biology* 45 (2001), pp. 771–95.

Benjamin, Walter. *The Arcades Project*. Trans. Howard Eiland and Kevin McLaughlin. Cambridge, MA: Belknap Press of Harvard University Press, 1999.

Bennett, Jim. "The Social History of the Microscope." *Journal of Microscopy* 155.3 (1989), pp. 267–80.

Bensaude-Vincent, Bernadette. "From Self-Organization to Self-Assembly: A New Materialism?" *History and Philosophy of the Life Sciences* 38.3 (2016), pp. 1–13.

_____. "Reconfiguring Nature through Syntheses: From Plastics to Biomimetics." In Bernadette Bensaude-Vincent and William R. Newman, eds., *The Artificial and the Natural: An Evolving Polarity*, pp. 294–312. Cambridge, MA: MIT Press, 2007.

_____, and Xavier Guchet. "Nanomachine: One Word for Three Different Paradigms." *Techné* 11.1 (2007), pp. 71–89.

Benton, E. "Vitalism in Nineteenth-Century Scientific Thought: A Typology and

Reassessment." *Studies in History and Philosophy of Science Part A* 5.1 (1974), pp. 17–48.

Berg, Howard C. "The Rotary Motor of Bacterial Flagella." *Annual Review of Biochemistry* 72 (2003), pp. 19–54.

Bergson, Henri. *Creative Evolution*. Trans. Arthur Mitchell. New York: Henry Holt, 1911.

———. *The Creative Mind: An Introduction to Metaphysics*. Trans. Mabelle L. Andison. New York: Philosophical Press, 1946.

Bermingham, Ann. *Learning to Draw: Studies in the Cultural History of a Polite and Useful Art*. New Haven, CT: Yale University Press, 2000.

Berryman, Sylvia. "The Imitation of Life in Ancient Greek Philosophy." In Jessica Riskin, ed., *Genesis Redux: Essays in the History and Philosophy of Artificial Life*, pp. 35–45. Chicago: University of Chicago Press, 2007.

Beyer, Andreas, and Guillaume Cassegrain, eds. *Mouvement/Bewegung: Über die dynamischen Potenziale der Kunst*. Berlin: Deutscher Kunstverlag, 2015.

Bitbol-Hespériès, Annie. "Cartesian Physiology." In Stephen Gaukroger, John Schuster, and John Sutton, eds., *Descartes' Natural Philosophy*, pp. 349–82. London: Routledge, 2000.

———. "De Vésale à Descartes: Le coeur, la vie." *Histoire des sciences médicales* 48.4 (2014), pp. 513–22.

Blay, Michel. *Reasoning with the Infinite: From the Closed World to the Mathematical Universe*. Trans. M. B. DeBevoise. Chicago: University of Chicago Press, 1998.

———. *La science du mouvement: De Galilée à Lagrange*. Paris: Belin, 2002.

Bleichmar, Daniela. *Visible Empire: Botanical Expeditions and Visual Culture in the Hispanic Enlightenment*. Chicago: University of Chicago Press, 2012.

Blumenberg, Hans. *Beschreibung des Menschen*. Ed. Manfred Sommer. Frankfurt am Main: Suhrkamp, 2006.

———. *Paradigms for a Metaphorology*. Trans. Robert Savage. Ithaca, NY: Cornell University Press, 2010.

Bognon-Küss, Cécilia, ed. "Organic, Organization, Organism: Essays in the History and Philosophy of Biology and Chemistry." Topical Collection, *History and Philosophy of the Life Sciences* 38.4 (2016).

Bolard, Jacques. "L'osmose et la vie selon Dutrochet." *BibNum* (2012), https://journals.openedition.org/bibnum/508.

Borelli, Giovanni Alfonso, *De motu animalium*. Rome: Angeli Bernabo, 1680–81.

———. *On the Movement of Animals*. Trans. Paul Maquet. Berlin: Springer, 1989.

Borisy, Gary, Rebecca Heald, Jonathon Howard, Carsten Janke, Andrea Musacchio, and Eva Nogales. "Microtubules: 50 Years on from the Discovery of Tubulin." *Nature Reviews: Molecular Cell Biology* 17.5 (2016), pp. 322–28.

Bracegirdle, Brian. *Beads of Glass: Leeuwenhoek and the Early Microscope.* Leiden: Museum Boerhaave, 1983.

———. *A History of Microtechnique: The Evolution of the Microtome and the Development of Tissue Preparation.* London: Heinemann, 1978.

Bradbury, Savile. *The Evolution of the Microscope.* Oxford: Pergamon, 1967.

Brandt, Frithiof. *Thomas Hobbes' Mechanical Conception of Nature.* Copenhagen: Levon & Munksgaard, 1927.

Braun, Marta. *Picturing Time: The Work of Etienne-Jules Marey (1830–1904).* Chicago: University of Chicago Press, 1992.

Braune, Wilhelm, and Otto Fischer. "Der Gang des Menschen: I. Theil: Versuche am unbelasteten und belasteten Menschen." *Abhandlungen der Königlich Sächsischen Gesellschaft der Wissenschaften: Mathematisch-physische Classe* 21 (1895), pp. 151–322.

Bredekamp, Horst. *Florentiner Fußball: Die Renaissance der Spiele.* Berlin: Wagenbach, 2001.

———. *The Lure of Antiquity and the Cult of the Machine: The Kunstkammer and the Evolution of Science, Art, and Technology.* Trans. Allison Brown. Princeton, NJ: M. Wiener, 1995.

———, Jochen Brüning, and Cornelia Weber, eds. *Theater der Natur und Kunst = Theatrum naturae et artis: Wunderkammern des Wissens.* Berlin: Henschel, 2000.

Breidbach, Olaf, Kerrin Klinger, and André Karliczek, eds. *Natur im Kasten: Lichtbild, Schattenriss, Umzeichnung und Naturselbstdruck um 1800.* Jena: Ernst-Haeckel-Haus, 2010.

Brookman, Philip, ed. *Eadweard Muybridge.* London: Tate, 2010.

Brouhard, Gary J. "Dynamic Instability 30 Years Later: Complexities in Microtubule Growth and Catastrophe." *Molecular Biology of the Cell* 26.7 (2015), pp. 1207–10.

Brown, Theodore M. "From Mechanism to Vitalism in Eighteenth-Century English Physiology." *Journal of the History of Biology* 7.2 (1974), pp. 179–216.

Brush, Stephen G. "A History of Random Processes: I. Brownian Movement from Brown to Perrin." *Archive for History of Exact Sciences* 5.1 (1968), pp. 1–36.

Buchan, Suzanne, ed. *Pervasive Animation.* New York: Routledge, 2013.

Buffon, Georges-Louis Leclerc. *Natural History: General and Particular.* Trans. William Smellie. 2nd ed. 9 vols. (Edinburgh: William Creech, 1780–85).

Buscaglia, Marino. "La pratique, la figure et les mots dans les *Mémoires* d'Abraham Trembley sur les polypes (1744), comme exemple de communication scientifique." In

Massimo Galuzzi, Gianni Micheli, and Maria Teresa Monti, eds., *Le forme della comunicazione scientifica*, pp. 313–46. Milan: Franco Angeli, 1998.

Bütschli, Otto. *Infusoria und System der Radiolaria*. Vol. 1.3 of *H. G. Bronns Klassen und Ordnungen des Tierreichs, wissenschaftlich dargestellt in Wort und Bild*. Leipzig: Winter'sche Verlagshandlung, 1887–89.

———. *Studien über die ersten Entwicklungsvorgänge der Eizelle, die Zelltheilung und die Conjugation der Infusorien*. Frankfurt am Main: C. Winter, 1876.

Byrne, Christopher. *Aristotle's Science of Matter and Motion*. Toronto: University of Toronto Press, 2018.

Cahan, David. *Helmholtz: A Life in Science*. Chicago: University of Chicago Press, 2018.

Cai, Dawen, Neha Kaul, Troy Lionberger, Diane Wiener, Kristen Verhey, and Edgar Meyhofer. "Recording Single Motor Proteins in the Cytoplasm of Mammalian Cells." *Methods in Enzymology* 475 (2010), pp. 81–107.

Canales, Jimena. "Photogenic Venus: The 'Cinematographic Turn' and Its Alternatives in Nineteenth-Century France." *Isis* 93.4 (2002), pp. 585–613.

———. *A Tenth of a Second: A History*. Chicago: University of Chicago Press, 2009.

Canguilhem, Georges. *Knowledge of Life*. Trans. Stefanos Geroulanos and Daniela Ginsburg. New York: Fordham University Press, 2008.

Carlet, Gaston. *Essai expérimental sur la locomotion humaine: Étude de la marche*. Paris: Martinet, 1872.

Cartwright, Lisa. *Screening the Body: Tracing Medicine's Visual Culture*. Minneapolis: University of Minnesota Press, 1995.

Castellani, Carlo. "Spermatozoan Biology from Leeuwenhoek to Spallanzani." *Journal of the History of Biology* 6.1 (1973), pp. 37–68.

Castiglioni, Arturo. "Galileo Galilei and His Influence on the Evolution of Medical Thought." *Bulletin of the History of Medicine* 12.2 (1942), pp. 226–41.

Chalfie, Martin. "GFP: Lighting Up Life." *Proceedings of the National Academy of Sciences* 106.25 (2009), pp. 10073–80.

Cheung, Tobias "What is an 'Organism'? On the Occurrence of a New Term and Its Conceptual Transformations 1680–1850." *History and Philosophy of the Life Sciences* 32.2–3 (2010), pp. 155–94.

Chevreton, Louise, and Frederic Vlès. "La cinématique de la segmentation de l'œuf et la chronophotographie du développement de l'oursin." *Comptes rendus hebdomadaires des séances de l'Académie des sciences* 149 (1909), pp. 806–809.

Cholodenko, Alan, ed. *The Illusion of Life: Essays on Animation*. Sydney: Power Publications, 1991.

Chong, Lisa, Elizabeth Culotta, and Andrew Sugden. "On the Move." *Science* 288.5463 (2000), p. 79.

Chouchourelou, Arieta, Alissa Golden, and Maggie Shiffrar. "What Does 'Biological Motion' Really Mean? Differentiating Visual Percepts of Human, Animal, and Nonbiological Motions." In Kerri L. Johnson and Maggie Shiffrar, eds., *People Watching: Social, Perceptual, and Neurophysiological Studies of Body Perception*, pp. 63–81. Oxford: Oxford University Press, 2013.

Chozinski, Tyler J., Lauren A. Gagnon, and Joshua C. Vaughan. "Twinkle, Twinkle Little Star: Photoswitchable Fluorophores for Super-Resolution Imaging." *FEBS Letters* 588.19 (2014), pp. 3603–12.

Churchill, Frederick B. "The Guts of the Matter: Infusoria from Ehrenberg to Bütschli, 1838–1876." *Journal of the History of Biology* 22.2 (1989), pp. 189–213.

———. "Introduction: Toward the History of Protozoology." *Journal of the History of Biology* 22.2 (1989), pp. 185–87.

Cimino, Guido, and François Duchesneau, eds. *Vitalisms from Haller to the Cell Theory*. Florence: Olschki, 1997.

Clark, William, Jan Golinski, and Simon Schaffer, eds. *The Sciences in Enlightened Europe*. Chicago: University of Chicago Press, 1999.

Coelho, Miguel, Nicola Maghelli, and Iva M. Tolić-Nørrelykke. "Single-Molecule Imaging *In Vivo*: The Dancing Building Blocks of the Cell." *Integrative Biology* 5.5 (2013), pp. 748–58.

Cohen, Barnett. "On Leeuwenhoek's Method of Seeing Bacteria." *Journal of Bacteriology* 34.3 (1937), pp. 343–46.

Cohen, Joel E. "Mathematics Is Biology's Next Microscope, Only Better; Biology Is Mathematics' Next Physics, Only Better." *PLoS Biology* 2.12 (2004), pp. 2017–23.

Cohen, Netta, and Jordan H. Boyle. "Swimming at Low Reynolds Number: A Beginners Guide to Undulatory Locomotion." *Contemporary Physics* 51.2 (2010), pp. 103–23.

Cohn, Ferdinand. *Contractile Gewebe im Pflanzenreiche*. Breslau: Josef Max, 1861.

———. *Zur Naturgeschichte des Protococcus pluvialis Kützing (Haematococcus pluvialis Flotow, Chlamidococcus versatilis A. Braun, Chlamidococcus pluvialis Flotow u. A. Braun)*. Bonn: Eduard Weber, 1850.

Cole, Francis Joseph. *Early Theories of Sexual Generation*. Oxford: Clarendon Press of Oxford University Press, 1930.

Collard, Patrick. *The Development of Microbiology.* Cambridge: Cambridge University Press, 1976.

Conklin, Edwin G. "Cell and Protoplasm Concepts: Historical Account." In Forest Ray Moulton, ed., *The Cell and Protoplasm*, pp. 6–19. Washington, DC: The Science Press, 1940.

Corbin, Alain, Jean-Jacques Courtine, and Georges Vigarello, eds. *Histoire du corps.* 3 vols. Paris: Seuil, 2005–2006.

Corcilius, Klaus. *Streben und Bewegen: Aristoteles' Theorie der animalischen Ortsbewegung.* Berlin: De Gruyter, 2008.

———, and Pavel Gregoric. "Aristotle's Model of Animal Motion." *Phronesis* 58.1 (2013), pp. 52–97.

Corliss, John O. *The Ciliated Protozoa: Characterization, Classification and Guide to the Literature.* 2nd ed. Oxford: Pergamon, 1979.

Corti, Bonaventura. *Osservazioni microscopiche sulla tremella e sulla circolazione del fluido in una pianta acquajuola.* Lucca: Giuseppe Rocchi, 1774.

Courty, Sébastien, Camilla Luccardini, Yohanns Bellaiche, Giovanni Cappello, and Maxime Dahan. "Tracking Individual Kinesin Motors in Living Cells Using Single Quantum-Dot Imaging." *Nano Letters* 6.7 (2006), pp. 1491–95.

Crafton, Donald. *Before Mickey: The Animated Film 1898–1928.* Cambridge, MA: MIT Press, 1984.

Crary, Jonathan. *Suspensions of Perception: Attention, Spectacle, and Modern Culture.* Cambridge, MA: MIT Press, 1999.

———. *Techniques of the Observer: On Vision and Modernity in the Nineteenth Century.* Cambridge, MA: MIT Press, 1991.

Crimp, Douglas. "Serra's Public Sculpture: Redefining Site Specificity." In Laura Rosenstock, ed., *Richard Serra: Sculpture*, pp. 40–56. New York: Museum of Modern Art, 1986.

Csiszar, Alex. *The Scientific Journal: Authorship and the Politics of Knowledge in the Nineteenth Century.* Chicago: University of Chicago Press, 2018.

Cunningham, Andrew. *The Anatomical Renaissance: The Resurrection of the Anatomical Projects of the Ancients.* Brookfield, VT: Scolar Press, 1997.

———. "The Pen and the Sword: Recovering the Disciplinary Identity of Physiology and Anatomy before 1800, II: Old Anatomy—The Sword." *Studies in History and Philosophy of Biological and Biomedical Sciences* 34.1 (2003), pp. 51–76.

Curry, Helen Anne, Nicholas Jardine, James Andrew Secord, and Emma C. Spary, eds. *Worlds of Natural History*. Cambridge: Cambridge University Press, 2018.

Curtis, Scott, ed. *Animation*. New Brunswick, NJ: Rutgers University Press, 2019.

———. "Die kinematographische Methode: Das 'bewegte Bild' und die Brownsche Bewegung." *Montage/av* 14.2 (2005), pp. 23–43.

———. *The Shape of Spectatorship: Art, Science, and Early Cinema in Germany*. New York: Columbia University Press, 2015.

Cutting, James E. "Gunnar Johansson, Events, and Biological Motion." In Kerri L. Johnson and Maggie Shiffrar, eds., *People Watching: Social, Perceptual, and Neurophysiological Studies of Body Perception*, pp. 11–21. Oxford: Oxford University Press, 2013.

Dagognet, François. *Étienne-Jules Marey: A Passion for the Trace*. Trans. Robert Galeta. New York: Zone Books, 1992.

Daston, Lorraine. "The History of Science and the History of Knowledge." *KNOW: A Journal on the Formation of Knowledge* 1.1 (2017), pp. 131–54.

———, and Peter Galison. *Objectivity*. New York: Zone Books, 2007.

Davidson, Michael W., and Richard N. Day, eds. *The Fluorescent Protein Revolution*. Boca Raton, FL: CRC Press, 2014.

Dawson, Virginia P. *Nature's Enigma: The Problem of the Polyp in the Letters of Bonnet, Trembley and Réaumur*. Philadelphia, PA: American Philosophical Society, 1987.

De Groot, Jean. *Aristotle's Empiricism: Experience and Mechanics in the Fourth Century BC*. Las Vegas, NV: Parmenides, 2014.

———. "*Dunamis* and the Science of Mechanics: Aristotle on Animal Motion." *Journal of the History of Philosophy* 46.1 (2008), pp. 43–67.

Deleuze, Gilles. *Bergsonism*. Trans. Hugh Tomlinson and Barbara Habberjam. New York: Zone Books, 1991.

Descartes, René. "Description of the Human Body." In *The Philosophical Writings of Descartes*. 3 vols. Trans. John Cottingham, Robert Stoothoff, and Dugald Murdoch, vol. 1, pp. 313–24. New York: Cambridge University Press, 1985.

Des Chene, Dennis. "Mechanisms of Life in the Seventeenth Century: Borelli, Perrault, Régis." *Studies in History and Philosophy of Biology and Biomedical Sciences* 36.2 (2005), pp. 245–60.

———. *Spirits and Clocks: Machine and Organism in Descartes*. Ithaca, NY: Cornell University Press, 2001.

Dickinson, Michael H., Claire T. Farley, Robert J. Full, M. A. R. Koehl, Rodger Kram, and Steven Lehman. "How Animals Move: An Integrative View." *Science* 288.5463 (2000), pp. 100–106.

Dienerowitz, Maria, Jamieson A. L. Howard, Steven D. Quinn, Frank Dienerowitz, and Mark C. Leake. "Single-Molecule FRET Dynamics of Molecular Motors in an ABEL Trap." *Methods* (online February 2021), n. p.

Dinsmore, Charles E., ed. *A History of Regeneration Research: Milestones in the Evolution of a Science.* Cambridge: Cambridge University Press, 1991.

Distelzweig, Peter M. "Descartes's Teleomechanics in Medical Context: Approaches to Integrating Mechanics and Teleology in Hieronymus Fabricius ab Aquapendente, William Harvey, and René Descartes." PhD diss., University of Pittsburgh, 2013.

———. "Fabricius's Galeno-Aristotelian Teleomechanics of Muscle." In Ohad Nachtomy and Justin E. H. Smith, eds., *The Life Sciences in Early Modern Philosophy*, pp. 65–84. Oxford: Oxford University Press, 2014.

Doane, Mary Ann. *The Emergence of Cinematic Time: Modernity, Contingency, and the Archive.* Cambridge, MA: Harvard University Press, 2002.

Dobell, Clifford. *Antony van Leeuwenhoek and His "Little Animals"; Being some Account of the Father of Protozoology and Bacteriology and His Multifarious Discoveries in these Disciplines.* New York: Harcourt, Brace, 1932.

Doherty, Meghan C. "Discovering the 'True Form:' Hooke's *Micrographia* and the Visual Vocabulary of Engraved Portraits." *Notes & Records of the Royal Society* 66.3 (2012), pp. 211–34.

Dolza, Luisa, and Hélène Vérin. "Figurer la mécanique: L'énigme des théâtres de machines de la Renaissance." *Revue d'histoire moderne et contemporaine* 51.2 (2004), pp. 7–37.

Drews, Gerhart. "The Roots of Microbiology and the Influence of Ferdinand Cohn on Microbiology of the 19th Century." *FEMS Microbiology Reviews* 24.3 (2000), pp. 25–49.

Duchesneau, François. *Genèse de la théorie cellulaire.* Collections analytiques. Montreal: Bellarmin, 1987.

———. *Les modèles du vivant de Descartes à Leibniz.* Paris: Vrin, 1998.

———. *La physiologie des Lumières: Empirisme, modèles et théories.* The Hague: Nijhoff, 1982.

———. "Territoires et frontières du vitalisme 1750–1850." In Guido Cimino and François Duchesneau, eds., *Vitalisms from Haller to the Cell Theory*, pp. 297–349. Florence: Olschki, 1997.

Dujardin, Félix. *Histoire naturelle des zoophytes: Infusoires, comprenant la physiologie et la classification de ces animaux, et la manière de les étudier à l'aide du microscope.* Paris: Roret, 1841.

———. *Nouveau manuel complet de l'observateur au microscope.* Paris: Librairie encyclopédique de Roret, 1843.

———. "Recherches sur les organismes inférieurs." *Annales des Sciences naturelles (zoologie)* 2, sér. 4 (1835), pp. 343–77.

Dunker, J. H. A. *Mikroskopische Blätter oder Beschreibungen und vergrößernde Abbildungen der kleinsten Werke Gottes: Zum Nutzen und Vergnügen für erwachsene Kinder und ungeübte Beobachter.* Brandenburg: Leichsche Buchhandlung, 1798.

Dupré, John, and Daniel J. Nicholson. "A Manifesto for a Processual Philosophy of Biology." In Daniel J. Nicholson and John Dupré, eds., *Everything Flows: Towards a Processual Philosophy of Biology*, pp. 1–78. Oxford: Oxford University Press, 2018.

Durantaye, Leland De La. "Kafka's Reality and Nabokov's Fantasy: On Dwarves, Saints, Beetles, Symbolism, and Genius." *Comparative Literature* 59.4 (2007), pp. 315–31.

During, Simon. *Modern Enchantments: The Cultural Power of Secular Magic.* Cambridge, MA: Harvard University Press, 2002.

Dutrochet, Henri. *L'agent immédiat du mouvement vital dévoilé dans sa nature et dans son mode d'action chez les végétaux et chez les animaux.* Paris: Baillière, 1826.

———. "Nouvelles observations sur l'endosmose et l'exosmose, et sur la cause de ce double phénomène." *Annales de chimie et de physique* 35 (1827), pp. 393–400.

———. *Recherches anatomiques et physiologiques sur la structure intime des animaux et des végétaux, et sur leur motilité.* Paris: Baillière, 1824.

Ecker, Alexander. *Zur Lehre vom Bau und Leben der contractilen Substanz der niedersten Thiere.* Basel: Schweighauser, 1846.

Ehrenberg, Christian Gottfried. *Die Infusionsthierchen als vollkommene Organismen: Ein Blick in das tiefere organische Leben der Natur.* Leipzig: Leopold Voss, 1838.

Elias, Norbert, and Eric Dunning, eds. *Quest for Excitement: Sport and Leisure in the Civilizing Process.* Oxford: Blackwell, 1986.

Elkins, James. "On Visual Desperation and the Bodies of Protozoa," *Representations* 40 (Autumn 1992), pp. 33–55.

Ellenbogen, Josh. "Camera and Mind." *Representations* 101.1 (2008), pp. 86–115.

———. "Educated Eyes and Impressed Images." *Art History* 33.3 (2010), pp. 490–511.

Elwick, James. "Herbert Spencer and the Disunity of the Social Organism." *History of Science* 41.1 (2003), pp. 35–72.

Erkoc, Pelin, Immihan C. Yasa, Hakan Ceylan, Oncay Yasa, Yunus Alapan, and Metin Sitti. "Mobile Microrobots for Active Therapeutic Delivery." *Advanced Therapeutics* 2.1 (2019), 1800064.

Espinas, Alfred. "L'organisation ou la machine vivante en Grèce, au IVe siècle avant J.-C." *Revue de Métaphysique et de Morale* 11.6 (1903), pp. 703–15.

Ewert, Malte. *Die Reichsanstalt für Film und Bild in Wissenschaft und Unterricht (1934–1945)*. Hamburg: Kovač, 1998.

Farley, John. *Gametes and Spores: Ideas about Sexual Reproduction, 1750–1914*. Baltimore, MD: Johns Hopkins University Press, 1982.

––––––. *The Spontaneous Generation Controversy from Descartes to Oparin*. Baltimore, MD: Johns Hopkins University Press, 1979.

Fauré-Fremiet, E. "L'oeuvre de Félix Dujardin et la notion de protoplasma." *Protoplasma* 23.1 (1935), pp. 250–69.

Fehr, Adrian N., Braulio Gutiérrez-Medina, Charles L. Asbury, and Steven M. Block. "On the Origin of Kinesin Limping." *Biophysical Journal* 97.6 (2009), pp. 1663–70.

Fehrenbach, Frank. "Kohäsion und Transgression: Zur Dialektik des lebendigen Bildes." In Ulrich Pfisterer and Anja Zimmermann, eds., *Transgressionen/Animationen: Das Kunstwerk als Lebewesen*, pp. 1–40. Berlin: Akademie, 2005.

––––––. *Leonardo da Vinci: Der Impetus der Bilder*. Berlin: Matthes & Seitz, 2019.

––––––. "Quasi animata forma: 'Living Art' in the Early Modern Period." In Marc Wellmann, ed., *BIOS: Konzepte des Lebens in der zeitgenössischen Skulptur*, pp. 30–39. Berlin: Wienand, 2012.

––––––. *Quasi vivo: Lebendigkeit in der italienischen Kunst der Frühen Neuzeit*. Berlin: De Gruyter, 2020.

––––––. "'Eine Zartheit am Rande unseres Sehvermögens': Bildwissenschaft und Lebendigkeit." *Kunst und Wissenschaft: Tendenzen, Probleme* 38.3 (2010), pp. 33–44.

Fernandez, Patricio A. "Reasoning and the Unity of Aristotle's Account of Animal Motion." In Brad Inwood, ed., *Oxford Studies in Ancient Philosophy*, vol. 47, pp. 151–203. Oxford: Oxford University Press, 2014.

Fili, Natalia. "Single-Molecule and Single-Particle Imaging of Molecular Motors *In Vitro* and *In Vivo*." In Christopher P. Toseland and Natalia Fili, eds., *Fluorescent Methods for Molecular Motors*, pp. 141–59. Basel: Springer, 2014.

Fischel, Angela, ed. *Instrumente des Sehens*. Berlin: De Gruyter, 2004.

––––––. "Optik und Utopie: Mikroskopische Bilder als Argument im 18. Jahrhundert." In

Horst Bredekamp and Pablo Schneider, eds., *Visuelle Argumentationen: Die Mysterien der Repräsentation und die Berechenbarkeit der Welt*, pp. 253–66. Munich: Fink, 2006.

Fleischer, Margot. *Anfänge europäischen Philosophierens: Heraklit, Parmenides, Platons Timaios.* Würzburg: Königshausen & Neumann, 2001.

Ford, Brian J. *Single Lens: The Story of the Simple Microscope.* New York: Harper & Row, 1985.

Foucault, Michel. *Discipline and Punish: The Birth of the Prison.* Trans. Alan Sheridan. New York: Vintage Books, 1979.

Fournier, Marian. *The Fabric of Life: Microscopy in the Seventeenth Century.* Baltimore, MD: Johns Hopkins University Press, 1996.

———. "Huygens' Designs for a Simple Microscope." *Annals of Science* 46.6 (1989), pp. 575–96.

Foxon, G. E. H. "Early Biological Film—The Work of R. G. Canti." *University Vision* 15 (1976), pp. 5–13

Frampton, Michael. *Embodiments of Will: Anatomical and Physiological Theories of Voluntary Animal Motion from Greek Antiquity to the Latin Middle Ages, 400 B.C.–A.D. 1300.* Saarbrücken: VDM, 2008.

Fraser, Mariam, Sarah Kember, and Celia Lury. "Inventive Life: Approaches to the New Vitalism." *Theory, Culture & Society* 22.1 (2005), pp. 1–14.

Freedberg, David. *The Eye of the Lynx: Galileo, His Friends, and the Beginnings of Modern Natural History.* Chicago: University of Chicago Press, 2002.

———. "Iconography between the History of Art and the History of Science: Art, Science, and the Case of the Urban Bee." In Caroline A. Jones and Peter Galison, eds., *Picturing Science, Producing Art*, pp. 272–96. New York: Routledge, 1998.

———, and Vittorio Gallese. "Motion, Emotion and Empathy in Esthetic Experience." *Trends in Cognitive Sciences* 11.5 (2007), pp. 197–203.

Freeland, Cynthia A. "Aristotle on Perception, Appetition, and Self-Motion." In Mary Louise Gill and James G. Lennox, eds., *Self-Motion: From Aristotle to Newton*, pp. 35–63. Princeton, NJ: Princeton University Press, 1994.

Friedl, Peter, Stefan Borgmann, and Eva-B. Bröcker. "Amoeboid Leukocyte Crawling through Extracellular Matrix: Lessons from the Dictyostelium Paradigm of Cell Movement." *Journal of Leukocyte Biology* 70.4 (2001), pp. 491–509.

Friedland, Alexander. " '. . . doch erscheint in seiner Denkschrift die Bedeutung des klinischen Films für den Unterricht allzustark betont': Zur Geschichte des

Medizinisch-kinematographischen Instituts der Charité 1923–1931." *Medizinhisto-risches Journal* 52.2–3 (2017), pp. 148–79.

Frixione, Eugenio. "Recurring Views on the Structure and Function of the Cytoskel-eton: A 300-Year Epic." *Cell Motility* 46.2 (2000), pp. 73–94.

Frizot, Michel, ed. *A New History of Photography*. Cologne: Könemann, 1998.

Frommolt, G. *Befruchtung, Furchung und erste Teilungen des Kanincheneies: Hochschulfilm C 23 des Instituts für Film und Bild*. Berlin: Reichsstelle für den Unterrichtsfilm, 1936.

———. "Die Befruchtung und Furchung des Kanincheneies im Film." *Zentralblatt für Gynäkologie* 58.1 (1934), pp. 7–12.

Furley, David J. "Self-Movers." In Mary Louise Gill and James G. Lennox, eds., *Self-Motion: From Aristotle to Newton*, pp. 3–14. Princeton, NJ: Princeton University Press, 1994.

Gadassik, Alla. "Assembling Movement: Scientific Motion Analysis and Studio Anima-tion Practice." *Discourse* 37.3 (2015), pp. 269–97.

Gaillard, Aurélia, Jean-Yves Goffi, Bernard Roukhomovsky, and Sophie Roux, eds. *L'Automate: Modèle, métaphore, machine, merveille*. Pessac: Presses universitaires de Bordeaux, 2013.

Gal, Ofer, and Yi Zheng, eds. *Motion and Knowledge in the Changing Early Modern World: Orbits, Routes and Vessels*. Dordrecht: Springer, 2014.

Galison, Peter. "Judgement against Objectivity." In Caroline A. Jones and Peter Galison, eds., *Picturing Science, Producing Art*, pp. 327–59. New York: Routledge, 1998.

Gaudreault, André, and Philippe Gauthier, eds. "Could Kinematography Be Anima-tion and Animation Kinematography?." Special issue, *animation: an interdisciplinary journal* 6.2 (2011).

Gaycken, Oliver. *Devices of Curiosity: Early Cinema and Popular Science*. Oxford: Oxford University Press, 2015.

———. "The Swarming of Life: Moving Images, Education, and Views through the Micro-scope." *Science in Context* 24.3 (2011), pp. 361–80.

Geison, Gerald L. "The Protoplasmic Theory of Life and the Vitalist-Mechanist Debate." *Isis* 60.3 (1969), pp. 273–92.

Gess, Nicola, Tina Hartmann, and Dominika Hens, eds. *Barocktheater als Spektakel: Mas-chine, Blick und Bewegung auf der Opernbühne des Ancien Régime*. Paderborn: Fink, 2015.

Geus, Armin. "Christian Gottfried Ehrenberg und 'Die Infusionsthierchen als vollkommene Organismen: Ein Blick in das tiefere organische Leben der Natur' (1838)." *Medizinhistorisches Journal* 22.2–3 (1987), pp. 228–45.

Giglioni, Guido. "Automata Compared: Boyle, Leibniz and the Debate on the Notion of Life and Mind." *British Journal for the History of Philosophy* 3.2 (1995), pp. 249–78.

Gill, Mary Louise. "Aristotle on Self-Motion." In Mary Louise Gill and James G. Lennox, eds., *Self-Motion: From Aristotle to Newton*, pp. 15–34. Princeton, NJ: Princeton University Press, 1994.

———, and James G. Lennox, eds. *Self-Motion: From Aristotle to Newton*. Princeton, NJ: Princeton University Press, 1994.

Gillgren, Peter. *Siting Michelangelo: Spectatorship, Site Specificity and Soundscape*. Lund: Nordic Academic Press, 2017.

Gleichen-Rußworm, Friedrich Wilhelm von. *Abhandlung ueber die Samen- und Infusionstierchen und ueber die Erzeugung: Nebst mikroskopischen Beobachtungen des Samens der Tiere und verschiedener Infusionen*. Nuremberg: Winterschmidt, 1778.

Gleichen, Wilhelm Friedrich Freiherr von, genannt Rußworm. *Das Neueste aus dem Reiche der Pflanzen, oder Mikroskopische Untersuchungen und Beobachtungen der geheimen Zeugungstheile der Pflanzen in ihren Blüten, und der in denselben befindlichen Insekten. . . .* Nuremberg: s.n., 1764.

Goldfuss, Georg August. *Ueber die Entwicklungsstufen des Thieres: Omne vivum ex ovo*. Nuremberg: Leonhard Schrag, 1817.

Goldschmidt, Richard. "Otto Bütschli 1848–1920." *Die Naturwissenschaften* 8.28 (1920), pp. 543–49.

Gombrich, Ernst H. "Moment and Movement in Art." In Gombrich, *The Image and the Eye: Further Studies in the Psychology of Pictorial Representation*, pp. 40–62. London: Phaidon, 1982.

Goodale, George L. "Protoplasm and Its History." *Botanical Gazette* 14.10 (1889), pp. 235–46.

Gorman, James. "This Tiny Robot Walks, Crawls, Jumps and Swims. But It Is Not Alive." *New York Times*, January 24, 2018, www.nytimes.com/2018/01/24/science/tiny-robot-medical.html.

Goss, Charles Mayo. "On Movement of Muscles by Galen of Pergamon." *American Journal of Anatomy* 123.1 (1968), pp. 1–25.

Gottdenker, Paula. "Three Clerics in Pursuit of 'Little Animals.'" *Clio Medica* 14.3–4 (1980), pp. 213–24.

Gotthelf, Allan, and James G. Lennox, eds. *Philosophical Issues in Aristotle's Biology*. Cambridge: Cambridge University Press, 1987.

Graaf, Reinier de. *De mulierum organis generationi inservientibus*. With an introduction by J. A. Dongen. Nieuwkoop: B. de Graaf, 1965.

Gräper, Ludwig. "Beobachtung von Wachstumsvorgängen an Reihenaufnahmen lebender Hühnerembryonen nebst Bemerkungen über vitale Färbung." *Morphologisches Jahrbuch* 39.3–4 (1911), pp. 304–27.

———. "Die frühe Entwicklung des Hühnchens nach Kinoaufnahmen des lebenden Embryo." *Anatomischer Anzeiger* 61 (1926), pp. 55–58.

———. "Die Gastrulation nach Zeitaufnahmen linear markierter Hühnerkeime." *Verhandlungen der Anatomischen Gesellschaft* (1937), pp. 6–12.

———. "Ludwig Ernst Gräper." In Bernd Wiederanders and Susanne Zimmermann, eds., *Buch der Docenten der Medicinischen Facultät zu Jena*, pp. 101–104. Golmsdorf: Jenzig, 2004.

———. "Die Methodik der stereokinematographischen Untersuchung des lebenden vitalgefärbten Hühnerembryos." *W. Roux' Archiv für Entwicklungsmechanik* 115 (1929), pp. 523–41.

———. "Die Primitiventwicklung des Hühnchens nach stereokinematographischen Untersuchungen, kontrolliert durch vitale Farbmarkierung und verglichen mit der Entwicklung anderer Wirbeltiere." *Wilhelm Roux' Archiv für Entwicklungsmechanik* 116 (1929), pp. 382–429.

———. "Die Primitiventwicklung des Hühnchens, verglichen mit der anderer Wirbeltiere, mit stereokinematographischen Demonstration." *Verhandlungen der Anatomischen Gesellschaft* (1928), pp. 90–96.

———. "Untersuchungen über die Herzbildung der Vögel." *Archiv für Entwicklungsmechanik* 24 (1907), pp. 375–410.

Greenleaf, William J., Michael T. Woodside, and Steven M. Block. "High-Resolution, Single-Molecule Measurements of Biomolecular Motion." *Annual Review of Biophysics and Biomolecular Structure* 36.1 (2007), pp. 171–90.

Gregoric, Pavel, and Martin Kuhar. "Aristotle's Physiology of Animal Motion: On *Neura* and Muscles." *Apeiron* 47.1 (2014), pp. 94–115.

Griendelius, Joannes Franciscus. *Micrographia nova: oder Neu-curieuse Beschreibung verschiedener kleiner Cörper, welche vermittelst eines absonderlichen von dem Author neuerfundenen Vergrösser-Glases verwunderlich gross vorgestellet werden*. Nuremberg: Johann Zieger, 1687.

Guerrini, Anita. *The Courtiers' Anatomists: Animals and Humans in Louis XIV's Paris*. Chicago: University of Chicago Press, 2015.

Gunning, Tom. "Animating the Instant: The Secret Symmetry between Animation and Photography." In Karen Beckman, ed., *Animating Film Theory*, pp. 37–53. Durham, NC: Duke University Press, 2014.

_____. "The Cinema of Attractions: Early Film, Its Spectator and the Avant-Garde." In Thomas Elsaesser, ed., *Early Cinema: Space, Frame, Narrative*, pp. 56–62. London: BFI Publishing, 1990.

_____. "Phantasmagoria and the Manufacturing of Illusions and Wonder: Towards a Cultural Optics of the Cinematic Apparatus." In André Gaudreault, Catherine Russell, and Pierre Veronneau, eds., *Le cinématographe, nouvelle technologie du XXe siècle = The Cinema, a New Technology for the 20th Century*, pp. 31–44. Lausanne: Payot, 2004.

Gustafson, Tryggve, and Lewis Wolpert. "The Cellular Basis of Morphogenesis and Sea Urchin Development." *International Review of Cytology* 15 (1963), pp. 139–214.

Hackney, D. D., J. D. Levitt, and J. Suhan. "Kinesin Undergoes a 9 S to 6 S Conformational Transition." *Journal of Biological Chemistry* 267.12 (1992), pp. 8696–8701.

Haeckel, Ernst. *Generelle Morphologie der Organismen: Allgemeine Grundzüge der organischen Formenwissenschaft, mechanisch begründet durch die von Charles Darwin reformirte Deszendenztheorie*. Berlin: Georg Reimer, 1866.

_____. *Das Protistenreich: Eine populäre Uebersicht über das Formengebiet der niedersten Lebewesen. Mit einem wissenschaftlichen Anhange: System der Protisten*. Leipzig: Ernst Günther, 1878.

_____. *Die Radiolarien: Eine Monographie*. Berlin: Georg Reimer, 1862.

Hall, Thomas Steele. *Ideas of Life and Matter: Studies in the History of General Physiology, 600 B.C.–1900 A.D.* 2 vols. Chicago: University of Chicago Press, 1969.

Haller, Albrecht von. *A Dissertation on the Sensible and Irritable Parts of Animals* (1752). Trans. with an introduction by Owsei Temkin, Baltimore: Johns Hopkins Press, 1936.

Hammond, John H. *The Camera Obscura: A Chronicle*. Bristol: Adam Hilger, 1981.

Harada, Yoshie, Katsuhiko Sakurada, Toshiaki Aoki, David D. Thomas, and Toshio Yanagida. "Mechanochemical Coupling in Actomyosin Energy Transduction Studied by *In Vitro* Movement Assay." *Journal of Molecular Biology* 216.1 (1990), pp. 49–68.

Harris, Henry. *The Birth of the Cell*. New Haven, CT: Yale University Press, 1999.

Harshey, Rasika M. "Bacterial Motility on a Surface: Many Ways to a Common Goal." *Annual Review of Microbiology* 57 (2003), pp. 249–73.

Harvey, William. *De motu locali animalium, 1627.* Ed. and trans. Gweneth Whitteridge. Cambridge: Cambridge University Press, 1959.

Hecht, Hermann. *Pre-Cinema History: An Encyclopaedia and Annotated Bibliography of the Moving Image before 1896.* London: Bowker Saur, 1993.

Heering, Peter. "The Enlightened Microscope: Re-enactment and Analysis of Projections with Eighteenth-Century Solar Microscopes." *British Journal for the History of Science* 41.3 (2008), pp. 345–67.

Herrmann, Hans-Christian von. "Motion Records: Die Bewegung im Zeitalter ihrer technischen Informierbarkeit." In Claudia Jeschke and Hans-Peter Bayerdörfer, eds., *Bewegung im Blick: Beiträge zu einer theaterwissenschaftlichen Bewegungsforschung,* pp. 100–12. Berlin: Vorwerk 8, 2000.

Heuser, J. E., and M. W. Kirschner. "Filament Organization Revealed in Platinum Replicas of Freeze-Dried Cytoskeletons." *Journal of Cell Biology* 86.1 (1980), pp. 212–34.

Hidalgo, Andrés, Edwin R. Chilvers, Charlotte Summers, and Leo Koenderman. "The Neutrophil Life Cycle." *Trends in Immunology* 40.7 (2019), pp. 584–97.

Hoffmann, E. T. A. "Master Flea." In *The Golden Pot and Other Tales.* Trans. with an introduction and notes by Ritchie Robertson. Oxford: Oxford University Press, 1992.

Hoogenraad Lab, "A Day in the Life of a Motor Protein," www.youtube.com /watch?v=tMKlPDBRJiE.

Hooke, Robert. "A Contrivance to make the Picture of any thing appear on a Wall, Cubboard, or within a Picture-frame, &c. in the midst of a Light room in the day-time; or in the Night-time in any room that is enlightned with a considerable number of Candles; devised and communicated by the Ingenious Mr. Hook, as follows." *Philosophical Transactions of the Royal Society of London* 3.38 (1668), pp. 741–43.

Hoppe, Brigitte. "Die Biologie der Mikroorganismen von F. J. Cohn (1828–1898): Entwicklung aus Forschungen über mikroskopische Pflanzen und Tiere." *Sudhoffs Archiv* 67.2 (1983), pp. 158–89.

Hopwood, Nick, Simon Schaffer, and James Secord. "Seriality and Scientific Objects in the Nineteenth Century." *History of Science* 48.3-4 (2010), pp. 251–85.

Horio, Tetsuya, and Hirokazu Hotani. "Visualization of the Dynamic Instability of Individual Microtubules by Dark-Field Microscopy." *Nature* 321.6070 (1986), pp. 605–607.

Hossfeld, Uwe. *"Im Dienst an Volk und Vaterland": Die Jenaer Universität der NS-Zeit.* Cologne: Böhlau, 2005.

_____, ed. *"Kämpferische Wissenschaft": Studien zur Universität Jena im Nationalsozialismus.* Cologne: Böhlau, 2003.

Howard, J., A. J. Hudspeth, and R. D. Vale, "Movement of Microtubules by Single Kinesin Molecules." *Nature* 342.6246 (1989), pp. 154–58.

Howard, Jonathon. "The Movement of Kinesin Along Microtubules." *Annual Review of Physiology* 58.1 (1996), pp. 703–29.

_____. "One Giant Step for Kinesin." *Nature* 365.6448 (1993), pp. 696–97.

Hu, Wenqi, Guo Zhan Lum, Massimo Mastrangeli, and Metin Sitti. "Small-Scale Soft-Bodied Robot with Multimodal Locomotion," *Nature* 554 (2018), pp. 81–85.

Huerta, Robert D. *Giants of Delft: Johannes Vermeer and the Natural Philosophers; The Parallel Search for Knowledge during the Age of Discovery.* Lewisburg, PA: Bucknell University Press, 2003.

Hunter, Matthew C. "'Mr. Hooke's Reflecting Box': Modeling the Projected Image in the Early Royal Society." *Huntington Library Quarterly* 78.2 (2015), pp. 301–28.

Huxley, Andrew F., and Rolf Niedergerke. "Structural Changes in Muscle During Contraction: Interference Microscopy of Living Muscle Fibres." *Nature* 173.4412 (1954), pp. 971–73.

Huxley, Hugh, and Jean Hanson. "Changes in the Cross-Striations of Muscle during Contraction and Stretch and Their Structural Interpretation." *Nature* 173.4412 (1954), pp. 973–76.

Inoué, S. "Video Image Processing Greatly Enhances Contrast, Quality, and Speed in Polarization-Based Microscopy." *Journal of Cell Biology* 89.2 (1981), pp. 346–56.

Isojima, Hiroshi, Ryota Iino, Yamato Niitani, Hiroyuki Noji, and Michio Tomishige. "Direct Observation of Intermediate States during the Stepping Motion of Kinesin-1." *Nature Chemical Biology* 12.4 (2016), pp. 290–97.

Jacobs, Natasha X. "From Unit to Unity: Protozoology, Cell Theory, and the New Concept of Life." *Journal of the History of Biology* 22.2 (1989), pp. 215–42.

Jaeger, Johannes, and Nick Monk. "Everything Flows: A Process Perspective on Life." *EMBO Reports* 16.9 (2015), pp. 1064–67.

Jahn, T. L., and E. C. Bovee. "Movement and Locomotion of Microorganisms." *Annual Review of Microbiology* 19 (1965), pp. 21–58.

Jantzen, Jörg. "Physiologische Theorien." In Francesco Moiso, Manfred Durner, and Jörg Jantzen, eds., *Wissenschaftshistorischer Bericht zu Schellings naturphilosophischen Schriften*, pp. 375–668. Stuttgart: Frommann-Holzboog, 1994.

Jaynes, Julian. "The Problem of Animate Motion in the Seventeenth Century." *Journal of the History of Ideas* 31.2 (1970), pp. 219–34.

Jeschke, Claudia, and Hans-Peter Bayerdörfer, eds. *Bewegung im Blick: Beiträge zur theaterwissenschaftlichen Bewegungsforschung.* Berlin: Vorwerk 8, 2000.

Jiang, Kevin. "To Boldly Go: New Microscope Captures 3-D Movies of Cells inside Living Organisms in Unprecedented Detail." Harvard Medical School News & Research, April 19, 2018, https://hms.harvard.edu/news/boldly-go.

Joblot, Louis. *Descriptions et usages de plusieurs nouveaux microscopes tant simples que composez, avec de nouvelles observations faites sur une multitude innombrable d'insectes et d'autres animaux de diverses espèces qui naissent dans les liqueurs préparées et dans celles qui ne le sont point.* Paris: Jacques Collombat, 1718.

Johansson, Gunnar. *Configurations in Event Perception: An Experimental Study.* Uppsala: Almqvist & Wiksell, 1950.

———. "Rigidity, Stability, and Motion in Perceptual Space: A Discussion of Some Mathematical Principles Operative in the Perception of Relative Motion and of Their Possible Function as Determinants of a Static Perceptual Space." *Acta Psychologica* 14 (1958), pp. 359–70.

———. "Spatio-Temporal Differentiation and Integration in Visual Motion Perception." *Psychological Research* 38.4 (1976), pp. 379–93.

———. "Vector Analysis in Visual Perception of Rolling Motion." *Psychologische Forschung* 36.4 (1974), pp. 311–19.

———. "Visual Perception of Biological Motion and a Model for Its Analysis." *Perception & Psychophysics* 14.2 (1973), pp. 201–11.

———. "Visual Vector Analysis and the Optic Sphere Theory." In Gunnar Jansson, Sten Sture Bergström, and William Epstein, eds., *Perceiving Events and Objects*, pp. 270–94. Hillsdale, NJ: Lawrence Erlbaum, 1994.

Johnson, Kerri L., and Maggie Shiffrar, eds. *People Watching: Social, Perceptual, and Neurophysiological Studies of Body Perception* Oxford: Oxford University Press, 2013.

Johnson, Mark H. "Biological Motion: A Perceptual Life Detector?" *Current Biology* 16.10 (2006), pp. R376–R377.

Jones, Caroline A., and Peter Galison, eds. *Picturing Science, Producing Art.* New York: Routledge, 1998.

Jones, Matthew L. *Reckoning with Matter: Calculating Machines, Innovating, and Thinking about Thinking from Pascal to Babbage.* Chicago: University of Chicago Press, 2016.

Joubin, Louis. "Notices biographiques, X: Félix Dujardin." *Archives de parasitologie* 4.1 (1901), pp. 5–57.

Kafka, Franz, "The Metamorphosis." Trans. Willa and Edwin Muir. In *The Complete Stories,* ed. Nahum N. Glatzer, pp. 89–139. New York: Schocken, 1971.

Kang, Minsoo. *Sublime Dreams of Living Machines: The Automaton in the European Imagination.* Cambridge, MA: Harvard University Press, 2011.

Kardel, Troels. "Nicolaus Steno's *New Myology* (1667): Rather than Muscle, the Motor Fibre Should Be Called Animal's Organ of Movement." *Nuncius* 23.1 (2008), pp. 37–64.

———. "Niels Stensen's Geometrical Theory of Muscle Contraction (1667): A Reappraisal." *Journal of Biomechanics* 23.10 (1990), pp. 953–65.

Kaulbach, Friedrich. *Der philosophische Begriff der Bewegung: Studien zu Aristoteles, Leibniz und Kant.* Cologne: Böhlau, 1965.

———, and Gerbert Meyer. S.v. "Bewegung." In Joachim Ritter, ed., *Historisches Wörterbuch der Philosophie.* Basel: Schwabe, 1971.

Keating, Jessica. *Animating Empire: Automata, the Holy Roman Empire, and the Early Modern World.* University Park: Pennsylvania State University Press, 2018.

Keller, Evelyn Fox. "Active Matter, Then and Now." *History and Philosophy of the Life Sciences* 38.3 (2016), pp. 1–11.

———. *Making Sense of Life: Explaining Biological Development with Models, Metaphors, and Machines.* Cambridge, MA: Harvard University Press, 2002.

Kemp, Martin, ed. *Leonardo on Painting.* New Haven, CT: Yale University Press, 2001.

———."Die Zeichen lesen: Zur graphischen Darstellung von physischer und mentaler Bewegung in den Manuskripten Leonardos." In Frank Fehrenbach, ed., *Leonardo da Vinci: Natur im Übergang,* pp. 207–27. Munich: Wilhelm Fink, 2002.

Kern, Stephen. *The Culture of Time and Space, 1880–1918.* Cambridge, MA: Harvard University Press, 1983.

Khatib, Oussama, Emel Demircan, Vincent De Sapio, Luis Sentis, Thor Besier, and Scott L. Delp. "Robotics-Based Synthesis of Human Motion." *Journal of Physiology-Paris* 103.3 (2009), pp. 211–19.

Kirschner, Marc, John Gerhart, and Tim Mitchison. "Molecular 'Vitalism.'" *Cell* 100.1 (2000), pp. 79–88.

Klemm, Friedrich. *Martin Frobenius Ledermüller: Aus der Zeit der Salon-Mikroskopie des Rokoko.* Schweidnitz: B. Köhn, 1928.

Klemm, Margot. *Ferdinand Julius Cohn 1828–1898: Pflanzenphysiologe, Mikrobiologe, Begründer der Bakteriologie*. Frankfurt am Main: Lang, 2003.

Koch, Johann Friedrich Wilhelm. *Mikrographie*. Vol. 1. Magdeburg: Keil, 1803.

Kolomeisky, Anatoly B. "Motor Proteins and Molecular Motors: How to Operate Machines at the Nanoscale." *Journal of Physics: Condensed Matter* 25.46 (2013), 463101.

Kopsch, Friedrich. "Beiträge zur Gastrulation beim Axolotl- und Froschei." *Verhandlungen der Anatomischen Gesellschaft Basel* 9 (1895), pp. 181–89.

——. "Ueber die Zellen-Bewegungen während des Gastrulationsprocesses an den Eiern vom Axolotl und vom braunen Grasfrosch." *Gesellschaft naturforschender Freunde Berlin* (1895), pp. 21–30.

Korten, Till, Bert Nitzsche, Chris Gell, Felix Ruhnow, Cécile Leduc, and Stefan Diez. "Fluorescence Imaging of Single Kinesin Motors on Immobilized Microtubules." In Erwin J. G. Peterman and Gijs J. L. Wuite, eds., *Single Molecule Analysis: Methods and Protocols*, pp. 121–37. Totowa, NJ: Humana Press, 2011.

Kraus, Andreas. "Der Beitrag Frankens zur Entwicklung der Wissenschaften (1550–1800)." In Andreas Kraus, ed., *Handbuch der Bayerischen Geschichte*, vol 3.1: *Geschichte Frankens bis zum Ausgang des 18. Jahrhunderts*, pp. 1054–1108. Munich: C. H. Beck, 1971.

Kringelbach, Hélène Neveu, and Jonathan Skinner, eds. *Dancing Cultures: Globalization, Tourism and Identity in the Anthropology of Dance*. New York: Berghahn, 2014.

Kuhl, Willi, and Hans Freska. *Die Entwicklung des Eies der weißen Maus*. Berlin: Reichsanstalt für Film und Bild, 1938.

Kühne, Wilhelm. *Untersuchungen über das Protoplasma und die Contractilität*. Leipzig: Engelmann, 1864.

Kützing, Friedrich. *Über die Verwandlung der Infusorien in niedere Algenformen*. Nordhausen: W. Köhne, 1844.

Kuriyama, Shigehisa. "'No Pain, No Gain' and the History of Presence." *Representations* 146.1 (2019), pp. 91–111.

Kwon, Miwon. *One Place after Another: Site-Specific Art and Local Identity*. Cambridge, MA: MIT Press, 2002.

Labarrière, Jean-Louis. *Langage, vie politique et mouvement des animaux: Études aristotéliciennes*. Paris: Vrin, 2004.

La Berge, Ann. "Medical Microscopy in Paris, 1830–1855." In Ann La Berge and Mordechai Feingold, eds., *French Medical Culture in the Nineteenth Century*, pp. 296–326. Leiden: Brill, 1994.

Lack, Elisabeth. *Kafkas bewegte Körper: Die Tagebücher und Briefe als Laboratorien von Bewegung.* Paderborn: Fink, 2009.

Laemmli, Whitney E. "The Living Record: Alan Lomax and the World Archive of Movement." *History of the Human Sciences* 31.5 (2018), pp. 23–51.

———. "Paper Dancers: Art as Information in Twentieth-Century America." *Information & Culture* 52.1 (2017), pp. 1–30.

Laks, André. *The Concept of Presocratic Philosophy: Its Origin, Development, and Significance.* Trans. Glenn W. Most. Princeton, NJ: Princeton University Press, 2018.

———, and Marwan Rashed, eds. *Aristote et le mouvement des animaux: Dix études sur le De motu animalium.* Villeneuve d'Ascq: Presses Universitaires de Septentrion, 2004.

———, and Claire Louguet, eds. *Qu'est-ce que la philosophie présocratique? What Is Presocratic Philosophy?* Villeneuve d'Ascq: Presses Universitaires de Septentrion, 2002.

Lamarck, Jean-Baptiste, *Zoological Philosophy: An Exposition with Regard to the Natural History of Animals.* Trans. Hugh Elliott. London: Macmillan, 1914.

Landecker, Hannah. "Creeping, Drinking, Dying: The Cinematic Portal and the Microscopic World of the Twentieth-Century Cell." *Science in Context* 24.3 (2011), pp. 381–416.

———. *Culturing Life: How Cells Became Technologies.* Cambridge, MA: Harvard University Press, 2007.

———. "The Life of Movement: From Microcinematography to Live-Cell Imaging." *Journal of Visual Culture* 11.3 (2012), pp. 378–99.

———, and Christopher Kelty. "A Theory of Animation: Cells, L-systems, and Film." *Grey Room* 17 (2004), pp. 30–63.

Lane, Nick. "The Unseen World: Reflections on Leeuwenhoek (1677) 'Concerning Little Animals.'" *Philosophical Transactions of the Royal Society B: Biological Sciences* 370 (2015), 20140344.

Lazarides, Elias, and Klaus Weber. "Actin Antibody: The Specific Visualization of Actin Filaments in Non-Muscle Cells." *Proceedings of the National Academy of Sciences* 71.6 (1974), pp. 2268–72.

Lechevalier, Hubert. "Louis Joblot and his Microscopes." *Bacteriological Reviews* 40.1 (1976), pp. 241–58.

———, and Morris Solotorovsky. *Three Centuries of Microbiology.* New York: McGraw-Hill, 1965.

Ledermüller, Martin Frobenius. *Ledermüllers . . . mikroskopische Gemüths- und Augen-*

Ergötzung: Bestehend, in Ein Hundert nach der Natur gezeichneten und mit Farben erleuchteten Kupfertafeln, Sammt deren Erklärung. Nuremberg: Christian de Launoy, 1761.

———. *Nachlese seiner mikroskopischen Gemüths- und Augenergötzung; 1. Sammlung. Bestehend in zehn illuminierten Kupfertafeln, sammt deren Erklärung; und einer getreuen Anweisung, wie man alle Arten Mikroskope, geschickt, leicht und nüzlich gebrauchen solle.* Nuremberg: Christian de Launoy, 1762.

———. *Physicalische Beobachtungen derer Saamenthiergens, durch die allerbesten Vergrößerungs-Gläser und bequemlichsten Microscope betrachtet; und mit einer unpartheyischen Untersuchung und Gegeneinanderhaltung derer Buffonischen und Leuwenhoeckischen Experimenten in einem Sendschreiben mit denen hierzu gehörigen Figuren und Kupfern einem Liebhaber der Natur-Kunde und Warheit mitgetheilet.* Nuremberg: George Peter Monath, 1756.

———. *Versuch einer gründlichen Vertheidigung derer Saamengethiergen: nebst einer kurzen Beschreibung der Leeuwenhoeckischen Mikroskopien und einem Entwurf zu einer vollständigern Geschichte des Sonnenmikroskops als der besten Rechtfertigung der Leeuwenhoeckischen Beobachtungen.* Nuremberg: George Peter Monath, 1758.

Leeuwenhoek-Commissie, ed. *Alle de brieven van Antoni van Leeuwenhoek / The Collected Letters of Antoni van Leeuwenhoek.* 15 vols. Amsterdam: Swets & Zeitlinger, 1939.

Leeuwenhoek, Antoni van. "An abstract of a Letter from Mr. Anthony Leevvenhoeck at Delft, dated Sep. 17. 1683, containing some Microscopical Observations, about Animals in the scurf of the Teeth, the substance call'd Worms in the Nose, the Cuticula consisting of Scales." *Philosophical Transactions of the Royal Society of London* 14.159 (1684), pp. 568–74.

———. "An Abstract of a Letter of Mr. Leeuwenhoeck Fellow of the R. Society, dated March 30th. 1685, to the R. S. Concerning Generation by an Insect." *Philosophical Transactions of the Royal Society of London* 15.174 (1685), pp. 1120–34.

———. "More Observations from Mr. Leewenhook, in a Letter of Sept. 7. 1674, sent to the Publisher." *Philosophical Transactions of the Royal Society of London* 9.108 (1674), pp. 178–82.

———. "Observations, communicated to the Publisher by Mr. Antony van Leewenhoeck, in a Dutch Letter of the 9th of Octob. 1676, here English'd: Concerning little Animals by him observed in Rain- Well- Sea- and Snow-water; as also in water wherein Pepper had lain infused." *Philosophical Transactions of the Royal Society of London* 12.133 (1677), pp. 821–31.

_____. "II. An Extract of a Letter from Mr. Leewenhoek, Dated the 10th of July, An. 1695. Containing Microscopical Observations on Eels, Mites, the Seeds of Figs, Strawberries, &c." *Philosophical Transactions of the Royal Society of London* 19.221 (1695), pp. 269–80.

_____. "IV. Part of a Letter from Mr Antony van Leeuwenhoek, F. R. S. concerning green Weeds growing in Water, and some Animalcula found about them." *Philosophical Transactions of the Royal Society of London* 23.283 (1703), pp. 1304–11.

_____. "XVI. A Letter from Mr. Anthouy van Leeuwenhoek, F. R. S. containing some further Microscopical Observations on the Animalcula found upon Duckweed, &c." *Philosophical Transactions of the Royal Society of London* 28.337 (1713), pp. 160–64.

Lefebvre, Thierry, Jacques Malthête, and Laurent Mannoni, eds. *Sur les pas de Marey: Science(s) et cinéma*. Paris: L'Harmattan, 2004.

Lennox, James G. *Aristotle's Philosophy of Biology: Studies in the Origins of Life Science*. Cambridge: Cambridge University Press, 2001.

_____. "The Complexity of Aristotle's Study of Animals." In Christopher Shields, ed., *The Oxford Handbook of Aristotle*, pp. 287–305. Oxford: Oxford University Press, 2012.

Leonhard, Karin. "Blut sehen." In Inge Hinterwaldner and Markus Buschhaus, eds., *The Picture's Image: Wissenschaftliche Visualisierung als Komposit*, pp. 104–28. Munich: Fink, 2006.

Leurechon, Jean. *Récréations mathématiques, composés de plusieurs problemes, plaisans & facetieux, d'arithmetique; geometrie, astrologie, optique, perspective, mechanique, chymie, & d'autres rares & curieux secrets: plusieurs desquels n'ont jamais esté imprimez*. Rouen: Charles Osmont, 1629.

Ley, Klaus, Carlo Laudanna, Myron I. Cybulsky, and Sussan Nourshargh. "Getting to the Site of Inflammation: The Leukocyte Adhesion Cascade Updated." *Nature Reviews Immunology* 7 (2007), pp. 678–89.

Leyssen, Sigrid, and Pirkko Rathgeber, eds. *Bilder animierter Bewegung/Images of Animate Movement*. Munich: Fink, 2013.

Li, Jinxing, Berta Esteban-Fernández de Ávila, Wei Gao, Liangfang Zhang, and Joseph Wang. "Micro/nanorobots for Biomedicine: Delivery, Surgery, Sensing, and Detoxification." *Science Robotics* 2.4 (2017), eaam6431.

Liesegang, Paul. *Wissenschaftliche Kinematographie: Einschließlich der Reihenphotographie*. Düsseldorf: Liesegang, 1920.

_____. *Zahlen und Quellen zur Geschichte der Projektionskunst und Kinematographie*. Berlin: Deutsches Drucks- und Verlagshaus, 1926.

Ligensa, Annemone, and Klaus Kreimeier, eds. *Film 1900: Technology, Perception, Culture.* New Barnet, UK: John Libbey, 2009.

Linnaeus, Carolus. *Systema naturae*, 12th ed., 3 vols. Stockholm: Lars Salvius, 1767.

Lippmann, Edmund O. von. *Urzeugung und Lebenskraft: Zur Geschichte dieser Probleme von den ältesten Zeiten an bis zu den Anfängen des 20. Jahrhunderts.* Berlin: J. Springer, 1933.

Liu, Daniel. "The Cell and Protoplasm as Container, Object, and Substance, 1835–1861." *Journal of the History of Biology* 50.4 (2017), pp. 889–925.

Liu, Tsung-Li, Srigokul Upadhyayula, Daniel E. Milkie, Ved Singh, Kai Wang, Ian A. Swinburne, Koshore R. Mosaliganti, et al. "Observing the Cell in Its Native State: Imaging Subcellular Dynamics in Multicellular Organisms." *Science* 360.6386 (2018), eaaq1392.

Loudcher, Jean-François. "Limites et perspectives de la notion de Technique du Corps de Marcel Mauss dans le domaine du sport." *STAPS* 91.1 (2011), pp. 9–27.

Ludwig, Heidrun. *Nürnberger naturgeschichtliche Malerei im 17. und 18. Jahrhundert.* Marburg an der Lahn: Basilisken-Presse, 1998.

Lüthy, C. H. "Atomism, Lynceus, and the Fate of Seventeenth-Century Microscopy." *Early Science and Medicine* 1.1 (1996), pp. 1–27.

Mahoney, Michael S. "Infinitesimals and Transcendent Relations: The Mathematics of Motion in the Late Seventeenth Century." In David C. Lindberg and Robert S. Westman, eds., *Reappraisals of the Scientific Revolution*, pp. 461–91. Cambridge: Cambridge University Press, 1990.

Malitsky, Joshua, and Oliver Gaycken, eds. "Science and Documentary." Special issue, *Journal of Visual Culture* 11.3 (2012).

Mallinckrodt, Rebekka von, ed. *Bewegtes Leben: Körpertechniken in der Frühen Neuzeit.* Wolfenbüttel: Herzog August Bibliothek, 2008.

———, and Angela Schattner, eds. *Sports and Physical Exercise in Early Modern Culture: New Perspectives on the History of Sports and Motion.* Abingdon, UK: Routledge, 2016.

Mann, Gunter. "Medizinisch-naturwissenschaftliche Buchillustration im 18. Jahrhundert in Deutschland." *Marburger Sitzungsberichte* 86.1–2 (1964), pp. 3–48.

Manning, Erin. "Grace Taking Form: Marey's Movement Machines." *Parallax* 14.1 (2008), pp. 82–91.

———. *Relationscapes: Movement, Art, Philosophy.* Cambridge, MA: MIT Press, 2009.

Mannoni, Laurent. *The Great Art of Light and Shadow: Archaeology of the Cinema.* Exeter: University of Exeter Press, 2000.

Marey, Étienne-Jules. *Du mouvement dans les fonctions de la vie: Leçons faites au Collège de France*. Paris: Germer Baillière, 1868.

———. *La machine animale: Locomotion terrestre et aérienne*. Paris: Germer Baillière, 1873.

———. *La méthode graphique dans les sciences expérimentales et principalement en physiologie et en médecine*. 2nd, rev. ed. Paris: G. Masson, 1885.

———. *Movement*. Trans. Eric Pritchard. New York: Appleton, 1895.

Martin, Benjamin. *The Young Gentleman and Lady's Philosophy: In a Continued Survey of the Works of Nature and Art by Way of Dialogue*. London: W. Owen, 1759.

Mather, George, and Sophie West. "Recognition of Animal Locomotion from Dynamic Point-Light Displays." *Perception* 22.7 (1993), pp. 759–66.

Matlin, Karl S., Jane Maienschein, and Manfred D. Laubichler, eds. *Visions of Cell Biology: Reflections Inspired by Cowdry's* General Cytology. Chicago: University of Chicago Press, 2018.

Mauss, Marcel. "Techniques of the Body." Trans. Ben Brewster. *Economy and Society* 2.1 (1973), pp. 70–88.

Mayer, Andreas. "The Physiological Circus: Knowing, Representing, and Training Horses in Motion in Nineteenth-Century France." *Representations* 111.1 (2010), pp. 88–120.

———. *The Science of Walking: Investigations into Locomotion in the Long Nineteenth Century*. Trans. Tilman Skowroneck and Robin Blanton. Chicago: University of Chicago Press, 2020.

Meli, Domenico Bertoloni. *Mechanism, Experiment, Disease: Marcello Malpighi and Seventeenth-Century Anatomy*. Baltimore: Johns Hopkins University Press, 2011.

Mendelsohn, Everett. "Cell Theory and the Development of General Physiology." *Archives internationales d'histoire des sciences* 16 (1963), pp. 419–29.

Métraux, Alexandre, and Andreas Mayer, eds. *Kunstmaschinen: Spielräume des Sehens zwischen Wissenschaft und Ästhetik*. Frankfurt am Main: Fischer, 2005.

Meyer, A. W. "The Discovery and Earliest Representations of Spermatozoa." *Bulletin of the Institute of the History of Medicine* 6.2 (1938), pp. 89–110.

Michaelis, Anthony R. *Research Films in Biology, Anthropology, Psychology, and Medicine*. New York: Academic Press, 1955.

Michaelis, Leonor. "Die vitale Färbung, eine Darstellungsmethode der Zellgranula." *Archiv für mikroskopische Anatomie* 55.1 (1899), pp. 558–75.

Mitchison, Tim, and Marc Kirschner. "Dynamic Instability of Microtubule Growth." *Nature* 312.5991 (1984), pp. 237–42.

Mohl, Hugo von. "Über die Saftbewegung im Innern der Zelle." *Botanische Zeitung* 4 (1846), pp. 73–87, 89–94.

Montalenti, Giuseppe, Paolo Rossi, Walter Bernardi, and Antonello La Vergata, eds. *Lazzaro Spallanzani e la biologia del Settecento: Teorie, esperimenti, istituzioni scientifiche.* Florence: Olschki, 1982.

Morel, Pierre-Marie. *De la matière à l'action: Aristote et le problème du vivant.* Paris: Vrin, 2007.

Morris, Geraldine, and Larraine Nicholas, eds. *Rethinking Dance History: Issues and Methodologies.* London: Routledge, 2018.

Mowll Mathews, Nancy, and Charles Musser, eds. *Moving Pictures: American Art and Early Film, 1880–1910.* Manchester, VT: Hudson Hills Press, 2005.

Müller, Gerhard H. "Martin Frobenius Ledermüllers Beziehungen zum Bayreuther Hof unter Markgraf Friedrich." *Jahrbuch für Fränkische Landesforschung* 38 (1978), pp. 171–80.

Müller, Otto Fridericus. *Animalcula infusoria fluviatilia et marina.* Copenhagen: Nicolai Möller, 1786.

Muybridge, Eadweard. *Descriptive Zoopraxography, or, The Science of Animal Locomotion Made Popular.* Philadelphia: University of Pennsylvania, 1893.

Myers, Natasha. "Animating Mechanism: Animations and the Propagation of Affect in the Lively Arts of Protein Modelling." *Science & Technology Studies* 19.2 (2006), p. 6–30.
_____. *Rendering Life Molecular: Models, Modelers, and Excitable Matter.* Durham, NC: Duke University Press, 2015.

Nagel, Alexander. *Medieval Modern: Art out of Time.* London: Thames & Hudson, 2012.

Naumann, Karl Wilhelm. *Ledermüller und von Gleichen-Russworm: Zwei deutsche Mikroskopisten der Zopfzeit.* Leipzig: Scholtze Nachf., 1926.

Needham, Joseph. *A History of Embryology.* Cambridge: Cambridge University Press, 1959.

Nicéron, Jean-François. *La perspective curieuse.* Paris: Langlois, 1652.

Nicholson, Daniel J. "The Machine Conception of the Organism in Development and Evolution: A Critical Analysis." *Studies in History and Philosophy of Science Part C: Studies in History and Philosophy of Biological and Biomedical Sciences* 48, Part B (2014), pp. 162–74.
_____. "Organisms ≠ Machines." *Studies in History and Philosophy of Science Part C: Studies in History and Philosophy of Biological and Biomedical Sciences* 44.4, Part B (2013), pp. 669–78.

Nicolson, Marjorie Hope. *Science and Imagination*. Ithaca, NY: Great Seal Books, 1956.

Niethammer, Verena. "Indoktrination oder Innovation? Der Unterrichtsfilm als neues Lehrmedium im Nationalsozialismus." *Journal of Educational Media, Memory, and Society* 81 (2016), pp. 30–60.

Nissen, Claus. *Die botanische Buchillustration*. 3 vols. Stuttgart: Hiersemann, 1966.

Nocek, Adam. *Molecular Capture: The Animation of Biology*. Minneapolis: University of Minnesota Press, 2021.

Nollet, Jean-Antoine. *Leçons de physique expérimentale* 3rd ed. Vol. 5. Paris: Guerin & Delatour, 1764.

Nosrati, Reza, Amine Driouchi, Christopher M. Yip, and David Sinton. "Two-Dimensional Slither Swimming of Sperm within a Micrometre of a Surface." *Nature Communications* 6 (2015), p. 8703.

Nussbaum, Martha Craven. *Aristotle's De motu animalium: Text with Translation, Commentary, and Interpretive Essays*. Princeton, NJ: Princeton University Press, 1985.

Nyhart, Lynn K. *Biology Takes Form: Animal Morphology and the German Universities, 1800–1900*. Chicago: University of Chicago Press, 1995.

Osler, Margaret J. "Eternal Truths and the Laws of Nature: The Theological Foundations of Descartes' Philosophy of Nature." *Journal of the History of Ideas* 46.3 (1985), pp. 349–62.

Ovid. *Ovid's Metamorphoses in Fifteen Books, Translated by the Most Eminent Hands*. London: Jacob Tonson, 1717.

Palmerino, Carla Rita, and J. J. M. H. Thijssen, eds. *The Reception of the Galilean Science of Motion in Seventeenth-Century Europe*. Dordrecht: Kluwer, 2004.

Pannabecker, John R. "Representing Mechanical Arts in Diderot's *Encyclopédie*." *Technology and Culture* 39.1 (1998), pp. 33–73.

Papapetros, Spyros. *On the Animation of the Inorganic: Art, Architecture, and the Extension of Life*. Chicago: University of Chicago Press, 2012.

Papenburg, Bettina, Liv Hausken, and Sigrid Schmitz, eds. "The Processes of Imaging/Imaging of Processes." Special section, *Catalyst: Feminism, Theory, Technoscience* 4.2 (2018).

Park, Hyokeun, Erdal Toprak, and Paul R. Selvin. "Single-Molecule Fluorescence to Study Molecular Motors." *Quarterly Reviews of Biophysics* 40.1 (2007), pp. 87–111.

Parnes, Ohad. "The Envisioning of Cells." *Science in Context* 13.1 (2000), pp. 71–92.

Pastre, Béatrice de, and Thierry Lefebvre. *Filmer la science, comprendre la vie: Le cinéma de Jean Comandon*. Paris: CNC, 2012.

Patin, Charles. *Travels thro' Germany, Bohemia, Swisserland, Holland, and other parts of Europe*. London: A. Swall and T. Child, 1696.

Perrig, Alexander. "Leonardo: Die Rekonstruktion menschlicher Bewegung." In Gottfried Schramm, ed., *Leonardo: Bewegung und Ruhe*, pp. 67–100. Freiburg: Rombach, 1997.

Peter, Karl. "Ludwig Gräper." *Anatomischer Anzeiger* (1937), pp. 300–18.

Peterman, Erwin J. G. "Kinesin's Gait Captured." *Nature Chemical Biology* 12.4 (2016), pp. 206–207.

Picken, Laurence. *The Organization of Cells and Other Organisms*. Oxford: Clarendon Press of Oxford University Press, 1960.

Pickstone, John V. "Globules and Coagula: Concepts of Tissue Formation in the Early Nineteenth Century." *Journal of the History of Medicine and Allied Sciences* 28.4 (1973), pp. 336–56.

——. "Vital Actions and Organic Physics: Henri Dutrochet and French Physiology during the 1820s." *Bulletin of the History of Medicine* 50.2 (1976), pp. 191–212.

Pollard, Thomas D., and Edward D. Korn. "Acanthamoeba Myosin: I. Isolation from Acanthamoeba Castellanii of an Enzyme Similar to Muscle Myosin." *Journal of Biological Chemistry* 248.13 (1973), pp. 4682–90.

Porter, J. R. "Antony van Leeuwenhoek: Tercentenary of his Discovery of Bacteria." *Bacteriological Reviews* 40.2 (1976), pp. 260–69.

Porter, Martin. *Windows of the Soul: The Art of Physiognomy in European Culture 1470–1780*. Oxford: Oxford University Press, 2005.

Porter, Roy, ed. *Eighteenth-Century Science*. Cambridge: Cambridge University Press, 2003.

Prodger, Phillip. "The Romance and Reality of the Horse in Motion." In Joyce Delimata, ed., *Actes du Colloque Marey/Muybridge, pionniers du cinéma: Rencontre Beaune/Stanford*, pp. 44–71. Beaune: Conseil régional de Bourgogne, 1995.

——. *Time Stands Still: Muybridge and the Instantaneous Photography Movement*. Oxford: Oxford University Press, 2003.

Purcell, E. M. "Life at Low Reynolds Number." *American Journal of Physics* 45.3 (1977), pp. 3–11.

Rabinbach, Anson. *The Human Motor: Energy, Fatigue, and the Origins of Modernity*. Berkeley: University of California Press, 1992.

Ratcliff, Marc J. *Genèse d'une découverte: La division des infusoires (1765–1766)*. Paris: Museum National Histoire Naturelle, 2016.

———. *The Quest for the Invisible: Microscopy in the Enlightenment*. Farnham, UK: Ashgate, 2009.

———. "Temporality, Sequential Iconography and Linearity in Figures: The Impact of the Discovery of Division in Infusoria." *History and Philosophy of the Life Sciences* 21.3 (1999), pp. 255–92.

———. "Wonders, Logic, and Microscopy in the Eighteenth Century: A History of the Rotifer" *Science in Context* 13.1 (2000), pp. 93–119.

Rathgeber, Pirkko. "Struktur- und Umrißmodelle als schematische Bilder der Bewegung." *Rheinsprung 11: Zeitschrift für Bildkritik* 2 (2011), pp. 130–53.

Rauber, August, and Friedrich Kopsch. *Lehrbuch und Atlas der Anatomie des Menschen*. Stuttgart: Thieme, 1870 et seq.

Reicke, Emil, ed. *Gottscheds Briefwechsel mit dem Nürnberger Naturforscher Martin Frobenius Ledermüller und dessen seltsame Lebensschicksale*. Leipzig: Scholtze, 1923.

Reill, Peter Hanns. *Vitalizing Nature in the Enlightenment*. Berkeley: University of California Press, 2005.

Reiß, Christian. "Shooting Chicken Embryos: The Making of Ludwig Gräper's Embryological Films 1911–1940." *Isis* 112.2 (2021), pp. 299–306.

Rescher, Nicholas. *Process Philosophy: A Survey of Basic Issues*. Pittsburgh, PA: University of Pittsburgh Press, 2000.

Reynolds, Andrew S. "Amoebae as Exemplary Cells: The Protean Nature of an Elementary Organism." *Journal of the History of Biology* 41 (2008), pp. 307–37.

———. "The Redoubtable Cell." *Studies in History and Philosophy of Science Part C: Studies in History and Philosophy of Biological and Biomedical Sciences* 41.3 (2010), pp. 194–201.

———. *The Third Lens: Metaphor and the Creation of Modern Cell Biology*. Chicago: University of Chicago Press, 2018.

Rheinberger, Hans-Jörg. *Toward a History of Epistemic Things: Synthesizing Proteins in the Test Tube*. Stanford, CA: Stanford University Press, 1997.

———, and Michael Hagner, eds. *Die Experimentalisierung des Lebens: Experimentalsysteme in den biologischen Wissenschaften 1850–1950*. Berlin: Akademie, 1993.

Richter, Wolfram. "Friedrich Kopsch als Histologe und Embryologe: Zur Erinnerung an den großen Berliner Anatomen aus Anlaß seines 30. Todestages am 24. Januar 1985." *Zeitschrift für mikroskopisch-anatomische Forschung* 99.1 (1985), pp. 1–13.

Ries, Julius. "Kinematographie der Befruchtung und Zellteilung." *Archiv für mikros-kopische Anatomie* 74 (1909), pp. 1–31.

Riskin, Jessica, ed. *Genesis Redux: Essays in the History and Philosophy of Artificial Life.* Chicago: University of Chicago Press, 2007.

———. *The Restless Clock: A History of the Centuries-Long Argument over What Makes Living Things Tick.* Chicago: University of Chicago Press, 2016.

Robertson, Lesley, Jantien Backer, Claud Biemans, Joop van Doorn, Klaas Krab, Willem Reijnders, Henk Smit, and Peter Willemsen. *Antoni van Leeuwenhoek: Master of the Minuscule.* Leiden: Brill, 2016.

Roche, Daniel. "Equestrian Culture in France from the Sixteenth to the Nineteenth Century." *Past and Present* 199.1 (2008), pp. 113–45.

Roger, Jacques. *The Life Sciences in Eighteenth-Century French Thought.* Ed. Keith R. Benson. Trans. Robert Ellrich. Stanford, CA: Stanford University Press, 1997.

———. "The Living World." In Roy Porter and G. S. Rousseau, eds., *The Ferment of Knowledge: Studies in the Historiography of Eighteenth-Century Science*, pp. 255–82. Cambridge: Cambridge University Press, 1980.

Rogers, Sheena. "Gunnar Johansson: A Practical Theorist: An Interview with William Epstein." In Gunnar Jansson, Sten Sture Bergström, and William Epstein, eds., *Perceiving Events and Objects*, pp. 3–25. Hillsdale, NJ: Lawrence Erlbaum, 1994.

Romeis, Benno. *Taschenbuch der mikroskopischen Technik.* Berlin: De Gruyter, 1943.

Rosenhof, August Johann Rösel von. *Der monatlich herausgegebenen Insecten-Belustigung.* Nuremberg: Privately published, 1736–52.

———. *Der monathlich-herausgegebenen Insecten-Belustigung dritter Theil, worinnen ausser verschiedenen, zu den in den beeden ersten Theilen enthaltenen Classen, gehörigen Insecten, auch mancheley Arten von acht neuen Classen nach ihrem Ursprung, Verwandlung und andern wunderbaren Eigenschaften aus eigener Erfahrung beschrieben, und in sauber illuminirten Kupfern, nach dem Leben abgebildet vorgestellet werden.* Nuremberg: Fleischmann, 1755.

Rossell, Deac. *Laterna magica — Magic Lantern.* Stuttgart: Füsslin, 2008.

———. "The Magic Lantern and Moving Images before 1800." *Barockberichte* 40–41 (2005), pp. 686–93.

———. "Nürnberg and the Bull's-Eye Magic Lantern." *New Magic Lantern Journal* 9.5 (2003), pp. 71–75.

Rothschild, Lynn J. "Protozoa, Protista, Protoctista: What's in a Name?" *Journal of the History of Biology* 22.2 (1989), pp. 277–305.

Rousseau, G. S., and Roy Porter, eds. *The Ferment of Knowledge: Studies in the Historiography of Eighteenth-Century Science*. Cambridge: Cambridge University Press, 1980.

Roux, Sophie. "Quelles machines pour quels animaux? Jacques Rohault, Claude Perrault, Giovanni Alfonso Borelli." In Aurélia Gaillard, Jean-Yves Goffi, Sophie Roux, and Bernard Roukhomovsky, eds., *L'automate: Machine, métaphore, modèle, merveille*, pp. 69–113. Bordeaux: Presses universitaires de Bordeaux, 2013.

Rudenko, Andrey, Luigi Palmieri, Michael Herman, Kris M. Kitani, Dariu M. Gavrila, and Kai O. Arras. "Human Motion Trajectory Prediction: A Survey." *International Journal of Robotics Research* 39.8 (2020), pp. 895–935.

Ruestow, Edward G. "Images and Ideas: Leeuwenhoek's Perception of the Spermatozoa." *Journal of the History of Biology* 16.2 (1983), pp. 185–224.

———. *The Microscope in the Dutch Republic: The Shaping of Discovery*. Cambridge: Cambridge University Press, 1996.

Rutherford, M. D., and Valerie A. Kuhlmeier, eds. *Social Perception: Detection and Interpretation of Animacy, Agency, and Intention*. Cambridge, MA: MIT Press, 2013.

Sachs, Joe. "Aristotle: Motion." *Internet Encyclopedia of Philosophy* (2005), www.iep.utm.edu/aris-mot.

Sander, Kathrin. *Organismus als Zellenstaat: Rudolf Virchows Körper-Staat-Metapher zwischen Medizin und Politik*. Herbolzheim: Centaurus, 2012.

Sapp, Jan. *Evolution by Association: A History of Symbiosis*. New York: Oxford University Press, 1994.

Sarasohn, Lisa T. "Motion and Morality: Pierre Gassendi, Thomas Hobbes and the Mechanical World-View." *Journal of the History of Ideas* 46.3 (1985), pp. 363–79.

Sawday, Jonathan. *Engines of the Imagination: Renaissance Culture and the Rise of the Machine*. London: Routledge, 2007.

Schickore, Jutta. "Fixierung mikroskopischer Beobachtungen: Zeichnung, Dauerpräparat, Mikrofotografie." In Peter Geimer, ed., *Ordnungen der Sichtbarkeit: Fotografie in Wissenschaft, Kunst und Technologie*, pp. 285–310. Frankfurt am Main: Suhrkamp, 2002.

Schiller, Joseph, and Tetty Schiller. *Henri Dutrochet (Henri du Trochet 1776–1847): Le matérialisme mécaniste et la physiologie générale*. Paris: Albert Blanchard, 1975.

Schleiden, Matthias Jacob. "Beiträge zur Phytogenesis." *Archiv für Anatomie, Physiologie und wissenschaftliche Medicin* (1838), pp. 137–76.

Schliwa, Manfred. "Kinesin: Walking or Limping?" *Nature Cell Biology* 5.12 (2003), pp. 1043–44.

Schmidgen, Henning. "Cinematography without Film: Architecture and Technologies of Visual Instruction in Biology around 1900." In Nancy Anderson and Michael R. Dietrich, eds., *The Educated Eye: Visual Culture and Pedagogy in the Life Sciences*, pp. 94–120. Hanover, NH: Dartmouth College Press, 2012.

———. "Lebensräder, Spektatorien, Zuckungstelegraphen: Zur Archäologie des physiologischen Blicks." In Helmar Schramm, ed., *Bühnen des Wissens: Interferenzen zwischen Wissenschaft und Kunst*, pp. 268–99. Berlin: Dahlem University Press, 2003.

Schnalke, Thomas. *Natur im Bild: Anatomie und Botanik in der Sammlung des Nürnberger Arztes Christoph Jacob Trew*. Exhibition catalogue. Erlangen: Universitätsverlag, 1995.

Schneider, Ulrich Johannes, ed. *Kulturen des Wissens im 18. Jahrhundert*. Berlin: De Gruyter, 2008.

Schultze, Max Johann Sigismund. "Über Muskelkörperchen und das was man eine Zelle zu nennen habe." *Archiv für Anatomie, Physiologie und wissenschaftliche Medicin* (1861), pp. 1–27.

———. *Das Protoplasma der Rhizopoden und der Pflanzenzellen: Ein Beitrag zur Theorie der Zelle*. Leipzig: Wilhelm Engelmann, 1863.

Schwann, Theodor. *Microscopical Researches into the Accordance in the Structure and Growth of Animals and Plants*. Trans. Henry Smith. London: Sydenham Society, 1847.

Schwemmer, Wilhelm. *Nürnberger Kunst im 18. Jahrhundert*. Nuremberg: Edelmann, 1974.

Sell, Robert. *Bewegung und Beugung des Sinns: Zur Poetologie des menschlichen Körpers in den Romanen Franz Kafkas*. Weimar: Metzler, 2002.

Shahsavan, Hamed, et al. "Bioinspired Underwater Locomotion of Light-Driven Liquid Crystal Gels." *Proceedings of the National Academy of Sciences* 117.10 (2020), pp. 5125–33.

Shapere, Alfred, and Frank Wilczek. "Geometry of Self-Propulsion at Low Reynolds Number." *Journal of Fluid Mechanics* 198 (1989), pp. 557–85.

Shapin, Steven, and Simon Schaffer. *Leviathan and the Air Pump: Hobbes, Boyle, and the Experimental Life*. Princeton, NJ: Princeton University Press, 1985.

Shapiro, Alan, ed. "Kepler, Optical Imagery, and the Camera Obscura." Special issue, *Early Science and Medicine* 13.3 (2008).

Sheetz, Michael P., and James A. Spudich. "Movement of Myosin-Coated Fluorescent Beads on Actin Cables *In Vitro*." *Nature* 303.5912 (1983), pp. 31–35.

Siebold, C. T. E. von. *Lehrbuch der vergleichenden Anatomie der wirbellosen Thiere*. Berlin: Veit, 1848.

Small, Edward S., and Eugene Levinson, "Toward a Theory of Animation." *Velvet Light Trap* 24 (1989), pp. 67–74.

Smith, Pamela H. *The Body of the Artisan: Art and Experience in the Scientific Revolution.* Chicago: University of Chicago Press, 2004.

_____, and Paula Findlen, eds. *Merchants and Marvels: Commerce, Science, and Art in Early Modern Europe.* London: Routledge, 2002.

Snyder, Joel. "Visualization and Visibility." In Caroline A. Jones and Peter Galison, eds., *Picturing Science, Producing Art*, pp. 379–97. New York: Routledge, 1998.

Snyder, Laura J. *Eye of the Beholder: Johannes Vermeer, Antoni van Leeuwenhoek, and the Reinvention of Seeing.* New York: Norton, 2015.

Solovev, Alexander A., Yongfeng Mei, Esteban Bermúdez Ureña, Gaoshan Huang, and Oliver G. Schmidt. "Catalytic Microtubular Jet Engines Self-Propelled by Accumulated Gas Bubbles." *Small* 5.14 (2009), pp. 1688–92.

Spallanzani, Lazzaro. *Opuscoli di fisica animale, e vegetabile. Volume secondo.* Modena: La Società Tipografica, 1776.

_____. *Tracts on the Nature of Animals and Vegetables.* Trans. John Graham Dalyell. Edinburgh: William Creech, 1799.

Spary, Emma C. *Utopia's Garden: French Natural History from Old Regime to Revolution.* Chicago: University of Chicago Press, 2000.

Spudich, James A., Stephen J. Kron, and Michael P. Sheetz. "Movement of Myosin-Coated Beads on Oriented Filaments Reconstituted from Purified Actin." *Nature* 315.6020 (1985), pp. 584–86.

Stacey, Jackie, and Lucy Suchman, eds. "Animation and Automation—The Liveliness and Labours of Bodies and Machines." Special issue, *Body & Society* 18.1 (2012).

Staden, Heinrich von. *Herophilus: The Art of Medicine in Early Alexandria. Edition, Translation and Essays.* Cambridge: Cambridge University Press, 1989.

Stadler, Ulrich. "Von Brillen, Lorgnetten, Fernrohren und Kuffischen Sonnenmikroskopen: Zum Gebrauch optischer Instrumente in Hoffmanns Erzählungen." In Hartmut Steinecke, ed., *E. T. A. Hoffmann: Deutsche Romantik im europäischen Kontext*, pp. 91–105. Berlin: Erich Schmidt, 1993.

Stafford, Barbara Maria. *Artful Science: Enlightenment Entertainment and the Eclipse of Visual Education.* Cambridge, MA: MIT Press, 1994.

_____. *Body Criticism: Imaging the Unseen in Enlightenment Art and Medicine.* Cambridge, MA: MIT Press, 1991.

———. "Images of Ambiguity: Eighteenth-Century Microscopy and the Neither/Nor." In David Philip Miller and Hanns Peter Reill, eds., *Visions of Empire: Voyages, Botany, and Representations of Nature*, pp. 230–57. Cambridge: Cambridge University Press, 1996.

———, and Frances Terpak. *Devices of Wonder: From the World in a Box to Images on a Screen.* Los Angeles: Getty Research Institute for the History of Art & the Humanities, 2001.

Steadman, Philip. *Vermeer's Camera: Uncovering the Truth behind the Masterpieces.* Oxford: Oxford University Press, 2001.

Steigerwald, Joan. *Experimenting at the Boundaries of Life: Organic Vitality in Germany around 1800.* Pittsburgh: University of Pittsburgh Press, 2019.

Steinke, Hubert. *Irritating Experiments: Haller's Concept and the European Controversy on Irritability and Sensibility, 1750–1790.* Amsterdam: Rodopi, 2005.

Stengers, Isabelle. "Beyond Conversation: The Risks of Peace." In Catherine Keller and Anne Daniell, eds., *Process and Difference: Between Cosmological and Poststructuralist Postmodernisms*, pp. 235–56. Albany: SUNY Press, 2002.

Stier, Friedrich. "Siedentopf, Henry Friedrich Wilhelm." *Neue Deutsche Biographie* 24 (2010), www.deutsche-biographie.de/pnd117629251.html#ndbcontent.

Stramer, Brian, and Graham A. Dunn. "Cells on Film—The Past and Future of Cinemicroscopy." *Journal of Cell Science* 128 (2015), pp. 9–13.

Sturm, Johann Christoph. *Collegium experimentale sive Curiosum.* Nuremberg: Endter, 1676.

Sutil, Nicolás Salazar. *Motion and Representation: The Language of Human Movement.* Cambridge, MA: MIT Press, 2015.

Sutter, Alexander. *Göttliche Maschinen: Die Automaten für Lebendiges bei Descartes, Leibniz, La Mettrie und Kant.* Frankfurt am Main: Athenäum, 1988.

Svoboda, Karel, Christoph F. Schmidt, Bruce J. Schnapp, and Steven M. Block. "Direct Observation of Kinesin Stepping by Optical Trapping Interferometry." *Nature* 365.6448 (1993), pp. 721–27.

Swain, Simon, ed. *Seeing the Face, Seeing the Soul: Polemon's Physiognomy from Classical Antiquity to Medieval Islam.* Oxford: Oxford University Press, 2007.

Swammerdam, Jan. *The Book of Nature; or the History of Insects.* London: C. G. Seyffert, 1758.

Szent-Györgyi, Albert, ed. *Studies from the Institute of Medical Chemistry, University of Szeged.* Vol. 1. *Myosin and Muscular Contraction.* Basel: Karger, 1942.

Taylor, D. L., and Y. L. Wang. "Molecular Cytochemistry: Incorporation of Fluorescently Labeled Actin into Living Cells." *Proceedings of the National Academy of Sciences of the United States of America* 75.2 (1978), pp. 857–61.

Taylor, Geoffrey Ingram. "Analysis of the Swimming of Microscopic Organisms." *Proceedings of the Royal Society of London. Series A. Mathematical and Physical Sciences* 209.1099 (1951), pp. 447–61.

Tipton, Jason A. *Philosophical Biology in Aristotle's Parts of Animals.* Heidelberg: Springer, 2014.

Toellner, Richard. *Albrecht von Haller: Über die Einheit im Denken des letzten Universalgelehrten.* Wiesbaden: Steiner, 1971.

Tosi, Virgilio. *Cinema before Cinema: The Origins of Scientific Cinematography.* Trans. Sergio Angelini. London: British Universities Film & Video Council, 2005.

Treiber-Merbach, Hildgard. *Gastrulation der Hühnerkeimscheibe: Erläuterungen zu dem gleichnamigen Film von Prof. Dr. med. Ludwig Gräper.* Berlin: Reichsanstalt für Film und Bild in Wissenschaft und Unterricht, 1939.

Treviranus, Ludolf Christian. *Beyträge zur Pflanzenphysiologie.* Göttingen: Heinrich Dieterich, 1811.

Troje, Nikolaus F. "What Is Biological Motion? Definition, Stimuli, Paradigms." In M. D. Rutherford and Valerie A. Kuhlmeier, eds., *Social Perception: Detection and Interpretation of Animacy, Agency, and Intention*, pp. 13–36. Cambridge, MA: MIT Press, 2013.

———, and Dorita H. F. Chang. "Shape-Independent Processing of Biological Motion." In Kerri L. Johnson and Maggie Shiffrar, eds., *People Watching: Social, Perceptual, and Neurophysiological Studies of Body Perception*, pp. 82–99. Oxford: Oxford University Press, 2013.

Truitt, E. R. *Medieval Robots: Mechanics, Magic, Nature, and Art.* Philadelphia: University of Pennsylvania Press, 2015.

Turner, Gerard L'Estrange. "Microscopical Communication." In Turner, *Essays on the History of the Microscope*, ch. 11. Oxford: Senecio, 1980.

Unger, Franz. *Anatomie und Physiologie der Pflanzen.* Pest: C. A. Hartleben, 1855.

———. *Botanische Briefe.* Vienna: Carl Gerold, 1852.

Vale, Ronald D. "How Lucky Can One Be? A Perspective from a Young Scientist at the Right Place at the Right Time." *Nature Medicine* 18.10 (2012), pp. 1486–88.

———, and Ronald A. Milligan. "The Way Things Move: Looking Under the Hood of Molecular Motor Proteins." *Science* 288.5463 (2000), pp. 88–95.

———, Thomas S. Reese, and Michael P. Sheetz. "Identification of a Novel Force-Generating Protein, Kinesin, Involved in Microtubule-Based Motility." *Cell* 42.1 (1985), pp. 39–50.

———, Bruce J. Schnapp, Thomas S. Reese, and Michael P. Sheetz. "Movement of Organelles along Filaments Dissociated from the Axoplasm of the Squid Giant Axon." *Cell* 40.20 (1985), pp. 449–54.

———, Bruce J. Schnapp, Thomas S. Reese, and Michael P. Sheetz. "Organelle, Bead, and Microtubule Translocations Promoted by Soluble Factors from the Squid Giant Axon." *Cell* 40.3 (1985), pp. 559–69.

Vartanian, Aram. "Trembley's Polyp, La Mettrie, and Eighteenth-Century Materialism." *Journal of the History of Ideas* 11.3 (1950), pp. 259–86.

Vermeir, Koen. "The Magic of the Magic Lantern (1660–1700): On Analogical Demonstration and the Visualization of the Invisible." *British Journal for the History of Science* 38.2 (2005), pp. 127–59.

Viel, Alain, Robert A. Lue, and John Liebler/XVIVO. "The Inner Life of the Cell." An edited three-and-a-half-minute version by David Bolinsky is available on Vimeo at https://vimeo.com/236991501.

Vienne, Florence. "Organic Molecules, Parasites, 'Urthiere': The Contested Nature of Spermatic Animalcules, 1749–1841." Trans. Kate Sturge. In Susanne Lettow, ed., *Gender, Race and Reproduction: Philosophy and the Early Life Sciences in Context*, pp. 45–63. Albany: SUNY Press, 2014.

Voskuhl, Adelheid. *Androids in the Enlightenment: Mechanics, Artisans, and Cultures of the Self.* Chicago: University of Chicago Press, 2013.

Wainwright, Milton. "An Alternative View of the Early History of Microbiology." *Advances in Applied Microbiology* 52 (2003), pp. 333–55.

Waldeyer, Anton. "Friedrich Kopsch." *Zeitschrift für mikroskopisch-anatomische Forschung* 61 (1955), pp. 155–58.

Wang, Yingxiao, John Y. J. Shyy, and Shu Chien. "Fluorescence Proteins, Live-Cell Imaging, and Mechanobiology: Seeing Is Believing." *Annual Review of Biomedical Engineering* 10.1 (2008), pp. 1–38.

Warburg, Aby. *Werke in einem Band.* Ed. Martin Treml, Sigrid Weigel, and Perdita Ladwig. Frankfurt am Main: Suhrkamp, 2010.

Weber, Wilhelm, and Eduard Weber. *Mechanics of the Human Walking Apparatus.* Trans. P. Maquet and R. Furlong. Berlin: Springer, 1992.

Wegner, Richard. "Christoph Jacob Trew (1695–1769): Ein Führer zur Blütezeit naturwissenschaftlicher Abbildungswerke in Nürnberg im 18. Jahrhundert." *Mitteilungen zur Geschichte der Medizin, der Naturwissenschaften und der Technik* 39 (1940), pp. 218–28.

Weibel, Douglas B., Piotr Garstecki, Declan Ryan, Willow R. DiLuzio, Michael Mayer, Jennifer E. Seto, and George M. Whitesides. "Microoxen: Microorganisms to Move Microscale Loads." *Proceedings of the National Academy of Sciences of the United States of America* 102.34 (2005), pp. 11963–67.

Weikard, Melchior A. *Biographie des Herrn Wilhelm Friedrich v. Gleichen genannt Rußworm Herrn auf Greifenstein, Bonnland und Ezelbach*. Frankfurt am Main: Hermann, 1783.

Weitmann, Pascal. *Technik als Kunst: Automaten in der griechisch-römischen Antike und deren Rezeption in der frühen Neuzeit als Ideal der Kunst oder Modell für Philosophie und Wissenschaft*. Tübingen: Wasmuth, 2011.

Wellmann, Janina, ed. "Cinematography, Seriality, and the Sciences." Special issue, *Science in Context* 24.3 (2011).

——. *The Form of Becoming: Embryology and the Epistemology of Rhythm, 1760–1830*. Trans. Kate Sturge. New York: Zone Books, 2017.

——. "Metamorphosis in Images: Insect Transformation from the End of the Seventeenth to the Beginning of the Nineteenth Century." In Gemma Anderson and John Dupré, eds., *Drawing Processes of Life: Molecules, Cells, Organisms*, pp. 237–71. Bristol: Intellect Press, 2022.

——. "Model and Movement: Studying Cell Movement in Early Morphogenesis, 1900 to the Present." *History and Philosophy of the Life Sciences* 40.3 (2018), pp. 1–25.

——. "Plastilin und Kreisel, Pinsel und Projektor: Julius Ries und die Materialität der seriellen Anschauung." In Gerhard Scholtz, ed., *Serie und Serialität: Konzepte und Analysen in Gestaltung und Wissenschaft*, pp. 77–93. Berlin: Reimer, 2017.

Wells, Paul. *Understanding Animation*. London: Routledge, 1998.

Whitehead, Alfred North. *Process and Reality: An Essay in Cosmology*. Cambridge: Cambridge University Press, 1929.

Wieland, Wolfgang. *Die aristotelische Physik*. Göttingen: Vandenhoeck & Ruprecht, 1962.

Williams, Drid. *Anthropology and the Dance: Ten Lectures*. Urbana: University of Illinois Press, 2004.

Wilson, Catherine. *The Invisible World: Early Modern Philosophy and the Invention of the Microscope*. Princeton, NJ: Princeton University Press, 1995.

Wilson, Edmund B. *The Cell in Development and Inheritance*. New York: Macmillan, 1896.

Woehlke, Günther, and Manfred Schliwa. "Walking on Two Heads: The Many Talents of Kinesin." *Nature Reviews Molecular Cell Biology* 1.1 (2000), pp. 50–58.

Wolfe, Charles T. "Do Organisms Have an Ontological Status?" *History and Philosophy of the Life Sciences* 32.2–3 (2010), pp. 195–231.

Wright, Thomas. *William Harvey: A Life in Circulation.* Oxford: Oxford University Press, 2013.

Wu, Zhiguang, Xiankun Lin, Xian Zou, Jianmin Sun, and Qiang He. "Biodegradable Protein-Based Rockets for Drug Transportation and Light-Triggered Release." *ACS Applied Materials & Interfaces* 7.1 (2015), pp. 250–55.

Xu, Haifeng, Mariana Medina-Sánchez, Veronika Magdanz, Lukas Schwarz, Franziska Hebenstreit, and Oliver G. Schmidt. "Sperm-Hybrid Micromotor for Targeted Drug Delivery." *ACS Nano* 12.1 (2018), pp. 327–37.

Yan, Xiaohui, Qi Zhou, Melissa Vincent, Yan Deng, Jiangfan Yu, Jianbin Xu, Tiantian Xu, et al. "Multifunctional Biohybrid Magnetite Microrobots for Imaging-Guided Therapy." *Science Robotics* 2.12 (2017), eaaq1155.

Yildiz, Ahmet, Michio Tomishige, Ronald D. Vale, and Paul R. Selvin. "Kinesin Walks Hand-Over-Hand." *Science* 303.5658 (2004), pp. 676–78.

——, Joseph N. Forkey, Sean A. McKinney, Taekjip Ha, Yale E. Goldman, and Paul R. Sevin. "Myosin V Walks Hand-Over-Hand: Single Fluorophore Imaging with 1.5-nm Localization." *Science* 300.5628 (2003), pp. 2061–65.

——, and Paul R. Selvin. "Fluorescence Imaging with One Nanometer Accuracy: Application to Molecular Motors." *Accounts of Chemical Research* 38.7 (2005), pp. 574–82.

——. "Kinesin: Walking, Crawling or Sliding Along?" *Trends in Cell Biology* 15.2 (2005), pp. 112–20.

Zahn, Johann. *Oculus artificialis teledioptricus sive telescopium.* 2nd ed. Nuremberg: Lochner, 1702.

Zammito, John H. *The Gestation of German Biology: Philosophy and Physiology from Stahl to Schelling.* Chicago: University of Chicago Press, 2017.

Zglinicki, Friedrich von. *Der Weg des Films: Die Geschichte der Kinematographie und ihrer Vorläufer.* Berlin: Privately published, 1956.

Zuidervaart, Huib J., and Douglas Anderson. "Antony van Leeuwenhoek's Microscopes and Other Scientific Instruments: New Information from the Delft Archives." *Annals of Science* 73.3 (2016), pp. 257–88.

Image Credits

1.1. Giovanni Alfonso Borelli, *De motu animalium* (Lugduni in Batavis: Gaesbeeck, Bou-testeyn, de Vivie, vander Aa, 1685), plate XI. Courtesy of SUB Göttingen: 8 ZOOL IV, 5315:1.

2.1. Étienne-Jules Marey, *Le mouvement* (Paris: G. Masson, 1894), p. 61, fig. 44. Courtesy of the Library, Max Planck Institute for the History of Science, Berlin.

2.2 Étienne-Jules Marey, *Le mouvement* (Paris: G. Masson, 1894), p. 21, fig. 12. Courtesy of the Library, Max Planck Institute for the History of Science, Berlin.

3.1. Gunnar Johansson, "Visual Perception of Biological Motion and a Model for Its Analysis," *Perception & Psychophysics* 14.2 (1973), p. 202, fig. 2. Reprinted by permission of Springer Nature.

4.1. Wenqi Hu et al., "Small-Scale Soft-Bodied Robot with Multimodal Locomotion," *Nature* 554 (2018), p. 83, fig. 2. Reprinted by permission of Springer Nature.

4.2. Yunus Alapan et al., "Microrobotics and Microorganisms: Biohybrid Autonomous Cellular Robots," *Annual Review of Control, Robotics, and Autonomous Systems* 2.1 (2019), p. 207, fig 1.

4.3. Klaus Ley et al., "Getting to the Site of Inflammation: The Leukocyte Adhesion Cascade Updated," *Nature Reviews Immunology* 7 (2007), p. 679, fig. 1. Reprinted by permission of Springer Nature.

5.1. Johann Swammerdam, *Bibel der Natur, worinnen die Insekten in gewisse Classen vert-heilt, sorgfältig beschrieben, zergliedert, in saubern Kupferstichen vorgestellt, mit vielen Anmerkungen über die Seltenheiten der Natur erleutert und zum Beweis der Allmacht des Schöpferrs angewendet werden* (Leipzig: Gleditsch, 1752), plate 31. Courtesy of SUB Göttingen: 2 ZOOL V, 6244.

5.2. Antoni van Leeuwenhoek, *Ondervindingen En Beschouwingen Der onsigbare geschapene waarheden, Waar in gehandelt wert Vande Schobbens inde Mond, de Lasarie, de Jeuking t'Kind met Vis-Schobbens, t' binnenste der Darmen, en de beweging derselve, als mede het Vet dat inde selve gevonden wert / Geschreven aande Wyt-beroemde Koninklyke Societeit In Engeland* (Leyden: van Gaesbeeck, 1684), p. 3. Courtesy of Staatsbibliothek zu Berlin — Preußischer Kulturbesitz: 50 MA 47150 R.

5.3. Antoni van Leeuwenhoek, "IV. Part of a Letter from Mr Antony van Leeuwenhoek, F. R. S. concerning green Weeds growing in Water, and some Animalcula found about them," *Philosophical Transactions of the Royal Society of London* 23.283 (1703), detail of fig. 8. Courtesy of SUB Göttingen: 4 PHYS. MATH V, 150.

5.4. Letter from Leeuwenhoek to Anthonie Heinsius, September 20, 1698, in *Ontledingen en Ontdekkingen van levene Dierkens in de Teel-deelen, van verscheyde Dieren, Vogelen en Visschen: van het Hout med der selver menigvuldige vaaten; van Hair, Vlees en Vis; Als mede van de grootemenigte der Dierkens in de Excrementen . . . vervat in verscheyde Brieven geschreven aan de Koniglijke Societeit in London* (Delft: van Krooneveld, 1702), p. 52. Courtesy of SUB Göttingen: 8 H NAT I, 9528:7.

5.5. Johann Zahn, *Oculus artificialis teledioptricus sive telescopium, ex abditis rerum naturalium et artificialium principiis protractum nova method, eaque solida explicatum ac comprimis e triplici fundamento physico seu naturali, methematico dioptrico et mechanic, seu practico stabilitum* (Herbipoli: Heyl, 1685), p. 728. Courtesy of SUB Göttingen: 4 PHYS III, 3255.

5.6. Martin Frobenius Ledermüller, *Ledermüllers . . . mikroskopische Gemüths- und Augen-Ergötzung: Bestehend, in Ein Hundert nach der Natur gezeichneten und mit Farben erleuchteten Kupfertafeln, Sammt deren Erklärung* (Nuremberg: Christian de Launoy, 1761), plate 51. Courtesy of SUB Göttingen: 4 BIBL UFF 596.

5.7. Martin Frobenius Ledermüller, *Ledermüllers . . . mikroskopische Gemüths- und Augen-Ergötzung: Bestehend, in Ein Hundert nach der Natur gezeichneten und mit Farben erleuchteten Kupfertafeln, Sammt deren Erklärung* (Nuremberg: Christian de Launoy, 1761), plate 71. Courtesy of SUB Göttingen: 4 BIBL UFF 596.

5.8 and 5.9. Martin Frobenius Ledermüller, *Nachlese Seiner Mikroskopischen Gemüths- und Augenergötzung; 1. Sammlung. Bestehend in zehn illuminierten Kupfertafeln, Sammt deren Erklärung: Und einer getreuen Anweisung, wie man alle Arten Mikroskope, geschickt, leicht und nüzlich gebrauchen solle* (Nuremberg: Christian de Launoy, 1762), plates 1 and 21. Courtesy of SUB Göttingen: 4 H NAT I, 9633:1.

5.10. Martin Frobenius Ledermüller, *Ledermüllers . . . mikroskopische Gemüths- und Augen-Ergötzung: Bestehend in Ein Hundert nach der Natur gezeichneeten und mit Farben erleuchteten Kupfertafeln, Sammt deren Erklärung* (Nuremberg: Christian de Launoy, 1761), plate 75. Courtesy of SUB Göttingen: 4 BIBL UFF 596.

5.11. Louis Joblot, *Descriptions et usages de plusieurs nouveaux microscopes tant simples que composez, avec de nouvelles observations faites sur une multitude innombrable d'insectes et d'autres animaux de diverses espèces qui naissent dans les liqueurs préparées et dans celles qui ne le sont point* (Paris: Jacques Collombat, 1718), plate 2. Courtesy of SUB Göttingen: 4 BIBL UFF 24.

6.1. August Johann Rösel von Rosenhof, *Der monathlich-herausgegebenen Insecten-Belustigung dritter Theil, worinnen ausser verschiedenen, zu den in den beeden ersten Theilen enthaltenen Classen, gehörigen Insecten, auch mancheley Arten von acht neuen Classen nach ihrem Ursprung, Verwandlung und andern wunderbaren Eigenschaften aus eigener Erfahrrung beschrieben, und in sauber illuminirten Kupfern, nach dem Leben abgebildet vorgestellet werden* (Nuremberg: Fleischmann, 1755), vol. 3, plate 101. Courtesy of SUB Göttingen: 8 ZOOL VI, 2200:3 RARA.

6.2. Félix Dujardin, "Recherches sur les organismes inférieurs," *Annales des Sciences naturelles (zoologie)*, 2, sér. 4 (1835), plate 10. Courtesy of SUB Göttingen: 8 H NAT I, 2065.

7.1. Ludwig Gräper, "Beobachtung von Wachstumsvorgängen an Reihenaufnahmen lebender Hühnerembryonen nebst Bemerkungen über vitale Färbung," *Morphologisches Jahrbuch* 39.3–4 (1911), plate 17, figs. 1–16. Courtesy of SUB Göttingen: 8 ZOOL II, 1000.

7.2. Ludwig Gräper, "Beobachtung von Wachstumsvorgängen an Reihenaufnahmen lebender Hühnerembryonen nebst Bemerkungen über vitale Färbung," *Morphologisches Jahrbuch* 39.3–4 (1911), fig. 4. Courtesy of SUB Göttingen: 8 ZOOL II, 1000.

7.3. Ludwig Gräper, "Die Primitiventwicklung des Hühnchens nach stereokinematographischen Untersuchungen, kontrolliert durch vitale Farbmarkierung und verglichen mit der Entwicklung anderer Wirbeltiere," *Wilhelm Roux' Archiv für Entwicklungsmechanik* 116 (1929), fig. 8. Courtesy of SUB Göttingen: 8 ZOOL III, 4733.

7.4. Ludwig Gräper, "Die Gastrulation nach Zeitaufnahmen linear markierter Hühnerkeime," *Verhandlungen der Anatomischen Gesellschaft* (1937), p. 7. Courtesy of SUB Göttingen: 8 ZOOL II, 210.

8.1. Animation of walking kinesin, created by BioVisions at Harvard University and the scientific animation XVIVO, Hartfort/CT, 2006. Image provided and reprinted by permission of Alain Viel, Harvard University.

8.2. Stills from "A Day in the Life of a Motor Protein," created by the Molecular

Neuroscience Laboratory led by Casper Hoogenraad at Utrecht University, 2013, www.youtube.com/watch?v=tMKlPDBRJ1E. Reproduced by permission of Casper Hoogenraad.

8.3. Erwin J. G. Peterman, "Kinesin's Gait Captured," *Nature Chemical Biology* 12.4 (2016), fig. 1. Reprinted by permission of Springer Nature.

Index

"Rockets," 88–89, 242 n.32.

Rosenhof, August Johann Rösel von: *Insect Amusements*, 111, 130, 257 n.3; portrayal of amoeba (little Proteus), 130–32, *131*, 144.

Rothschild, Lynn, 135.

Rotifera, 99–101, 140, 243 n.7, 257 n.7.

Roux, Wilhelm, 160.

Royal Society, 95, 244 n.15; *Philosophical Transactions*, 92, 107.

SACHS, JOE, translation of Aristotle, 30, 221 n.4.

Samsa, Gregor (fictional character), 9–11, 13, 211, 215 n.1.

Sarcode, 146, 147, 148, 177. *See also* Protoplasm.

Sarcodina, 135. *See also* Amoebae.

Saturn's rings, 107.

Saussure, Horace Bénédicte de, 128.

Schleiden, Matthias, 136.

Schultze, Max, 149–50.

Schultze, Oskar, 160.

Schultz-Schultzenstein, Carl Heinrich, 142–43, 263 n.59.

Schwann, Theodor, 140; cell theory of, 136–37.

Science (journal), 19.

Scientific Revolution, 16.

Sea urchin morphogenesis, 171, 268 n.29.

Self-motion, 34–35, 51, 223 n.15. *See also* Voluntary motion.

Self-motivation, 71.

Sensation: Galen on, 42–43, 226–27 n.45; separated from motion, 50–51.

Sensory perception, 63–64, 247 n.45. *See also* Perception; Visual perception of biological motion.

Serial photography, 156–59, 160, 162.

Sheetz, Michael, 181, 182–83.

Shiffrar, Maggie, 238 n.14.

Shrimp, 102–103.

Siebold, Carl Theodor von, 134, 146.

Siedentopf, Henry, 162.

Sinews, 41, 42, 43. *See also* Neura.

Single-molecule fluorescence, 198, 200–201.

Siting, 21–23.

Sitti, Metin, 217 n.7.

Sliding filament hypothesis, 178, 181, 273 n.13.

Slime molds, 148.

"Snail-o-bots," 17, 217 n.7.

Snyder, Joel, 234 n.30.

Soft-matter physics, 240 n.10.

Soft robots, 77–79, 82, 208; locomotion of, 76.

Solar eclipses, 106.

Solar microscope, 25, 103, 110–15, 249 n.50, 251 n.67; sperm observed through, 122; Ledermüller's staging and descriptions, 115–21, *116–17*, 253 n.74; used by Dutrochet, 143; Zahn's "Lucerna megalographica, 109–110, *109*.

Soul: anatomy and, 41; Aristotle's concept of, 31–32, 34–35, 38, 41, 44; Galen's concept of, 42–43, 226–27 n.45; Haller's concept of, 50–51; in Harvey's view of movement, 44; location in the brain, 42, 43; motion bestowed by, 31–32, 34–35, 38, 44, 48; Plato's concept of, 43; presence before birth, 98.

Space and time, 40, 159, 173, 186; in stereoscopy, 164–68, 269 n.39; tracking, 199, 201, 203, 206.

Spallanzani, Lazzaro, 123–24, 128, 140; *Tracts on the Nature of Animals and Vegetables*, 123.

Sperm: as animals, 97–98, 121–24, 246 n.30; Buffon's view of, 121; observed with solar microscope, 122; "spermbot," 88; studied by Huygens, 249 n.51; swimming of, 20, 87–88, 97–98, 122, 123, 217 n.7.

Sphygmograph, 55.

Spirogyra, 243 n.7.

Spirulina, 87.

Spontaneous generation, 97, 128.

Sporozoans, 135.

Spudich, James, 181.

Squid axons, 181–82, 183.

Stahl, Georg Ernst, 51.

Stanford, Leland, 58.

Staub, Bruno, 178.

Stensen, Niels, 229 n.66.

Stereoscopy, 164–67, 269 n.39; stereo-cinematography, 164–66, *166*, 269 n.39.

Stiegler, Bernard, 217 n.9.

Stimulus-response, 17, 50.

Strathern, Marilyn, 272 n.5.

Sturm, Johann Christoph, 108; *Collegium experimentale sive Curiosum*, 108.

Subcellular motion, 80–81, 173, 175, 179, 201. *See also* Molecular motion.

Zone Books series design by Bruce Mau

Image placement and production by Julie Fry

Typesetting by Meighan Gale

Printed and bound by Maple Press